Cosmic Threats

Cosmic Threats: A Planetary Perspective is a continuing contribution to a previously published trilogy, detailed below, which seeks to reconnoitre the salient issues involved in building a durable world peace.

Engaging the Cosmos: ***Astronomy, Philosophy and Faith*** **(2006)**
"Neville Brown makes it natural that astronomy, philosophy and faith in their widest interpretation should be discussed together. Each reinforces the other in the development of the book. It is rare to find an author who is equally comfortable with modern astrophysics as with the motivations of the ancient world. If you want to know who is or was who in these fields you will find a discussion of them in this book. The issues are not left to ideals. We are brought to face issues of modern conscience and responsibility." *Professor Yvonne Elsworth, FRS, School of Physics and Astronomy, The University of Birmingham*

The Geography of Human Conflict: ***Approaches to Survival*** **(2009)**
"'Geopolitics' is a much misunderstood term that is over-used in modern politics. Neville Brown puts geopolitics back where it belongs – as a scientific and particular way of interpreting world politics that offers both explanation and meta-prediction. And he does so in a way that is both delightful and impressive. He offers a sweep of history, culture and science that is as breathtaking as it is riveting. If students of global politics are frightened of being changed simply by reading one book, they should stay away from this one. It will stretch and convert them in a single reading." *Professor Michael Clarke, Director, Royal United Services Institute for Defence and Security Studies*

The Bounds of Liberalism: ***The Fragility of Freedom*** **(2013)**
"Although the lynchpin of this reflection is social liberalism it is not an uncritical one: his historical awareness is too acute for such an easy approach to be acceptable. Rather, there is a continuous probing and testing of both the nature of the pressures to which polities across the world are subject and the governmental solutions which, since the dawn of human political history, have been proffered. There is emphasis on linkage: science, philosophy and history are presented as integral to political understandings, not as separate contributors. Equally characteristic is its intellectual restlessness and refusal to accept uncritically common conceptual assumptions be these with regard to parliamentary democracy or happiness and fulfilment." *Brian Salter, Professor of Politics, Department of Political Economy, King's College London*

Cosmic Threats

A PLANETARY RESPONSE

NEVILLE BROWN

SUSSEX
ACADEMIC
PRESS
Brighton • Chicago • Toronto

2 4 6 8 10 9 7 5 3 1

First published in Great Britain in 2017 by
SUSSEX ACADEMIC PRESS
PO Box 139, Eastbourne BN24 9BP

and in the United States of America by
SUSSEX ACADEMIC PRESS
Independent Publishers Group
814 N Franklin St, Chicago, IL 60610, USA

British Library Cataloguing in Publication Data
A CIP catalogue record for this book is available from the British Library.

Library of Congress Cataloging-in-Publication Data
Names: Brown, Neville, author.
Title: Cosmic threats : a planetary response / Neville Brown.
Description: Brighton ; Portland : Sussex Academic Press, 2018. | Includes bibliographical references and index.
Identifiers: LCCN 2017014798| ISBN 9781845197704 (hardback : acid-free paper) | ISBN 9781845197711 (paperback : acid-free paper)
Subjects: LCSH: Emergency management—International cooperation. | Disaster relief—International cooperation. | Natural disasters—Government policy. | Collisions (Astrophysics)—Government policy. | Security, International. | Defensive (Military science)
Classification: LCC HV551.2 .B77 2018 | DDC 363.34/526—dc23
LC record available at https://lccn.loc.gov/2017014798

Typeset & designed by Sussex Academic Press, Brighton & Eastbourne.

Contents

Prefatory Statement

This preface is involved more directly with the analysis to follow than are most. The reason is that, if the analysis in the body of the text can be broadly accepted, the implications can be read at two levels. The one is essentially that we should prepare, militarily and politically, against a random though menacing form of overhead threat. The other is that this preparation needs be underlaid by a thorough review of our management of planet Earth, as set within the cosmos as a whole. In other words, protection against Near Earth Objects (NEOs) from Outer Space (comets, asteroids, meteors) might beneficially be made focal to addressing all Nature's threats to our survival. So how readily might incoming asteroids and comets be destroyed or diverted?

This volume could well be my last book on contemporary strategic studies. My debut, *Strategic Mobility*, was embarked on as I became a member of the first research team to form up in-house at the newly found Institute for Strategic Studies (ISS), this in the year or two prior to the Cuba missile crisis of 1962. There were four of us research staffers. They were Alastair Buchan, Leonard Beaton, Philip Windsor and myself. The Eternal Reaper has since claimed my three colleagues, each when still in his impressive prime. In 1964, the ISS was to become the International Institute for Strategic Studies (IISS), a transformational identity development.

The late Denis Healey, Britain's Minister of Defence from 1964, was to be widely regarded as "the best Prime Minister Britain never had". He had been decisively influential in getting the ISS launched. So was he soon to be in having the Royal United Services Institute (RUSI) revamped to become an institution addressing synoptically Britain's security perspectives. Along with the IISS, the RUSI has since secured for London something of a pole position in strategic studies, as this newly identified subject area has proliferated worldwide.

Terrestrial Climate Change

Yet there have been shortfalls across the scene, shortfalls liable to bear directly on the outlook for planetary security. Thus inadequate account has so far been taken of the near term implications of global climate change.

On this score, however, primary responsibility has rested historically not with the strategists but with the meteorological fraternity. During the 1810s, Joseph Fraunhofer, a German optics engineer, confirmed that every gas has its unique spread of absorption lines within the spectrum of electromagnetic wavelengths. In 1829, Jean-Baptiste Fourier (one of several eminent mathematicians Napoleon had gathered about himself) demonstrated how, rather like a plate of glass over a box, carbon dioxide increasing in the air through industrialization can act as a blanket to push up temperatures within. For "box" now read "greenhouse".

In 1859 and 1906 respectively, two distinguished physical scientists, John Tyndall and Svante Arrhenius, were to demonstrate definitively the greenhouse properties, as expressed within the atmosphere, of carbon dioxide and of water vapour, the former lately proving especially prone to increase incrementally over time. But neither their insights nor other individual contributions made much impression on the stolid phalanxes of weather watchers.

Not until the Intergovernmental Panel on Climate Change was founded in 1990 could one say that the professional community was ready to engage in discourse.[1] What still is badly lacking is an appreciation of the likely effect of on-going climate change on particular regions already troubled by other factors. The contemporary Middle East comes horrendously to mind.

Aesthetics and Toxicity

Also to dwell on further is how the aesthetic and salubrious aspects of environment management can be affected by induced climate change or other man-made influences. Such Russian writers as Stalin and Solzhenitsyn have seen something to extoll in the customarily hard winters of inner Russia. A more ice-free Northern Sea Route from Archangel to the Behring Straits will not bring that "something" back. Likewise, the fact that the British climate has been quite equable yet stimulating virtually throughout the last two millennia has been seen as conducive to political moderation. Contrariwise, many people across the modern world are victims, physically and psychologically, of pell-mell urban growth further aggravated by global warming. Witness those billion Chinese who repair to their rural roots awhile every Chinese New Year, many to escape from a soulless, oppressive and toxic urban ambience.

Action–Reaction

To my mind, Action–Reaction is another security tendency which is being badly underestimated by moderate opinion, both in public affairs and in academe. Yet faced with a shrinking world, any social grouping is liable to become not more open but more inward looking. And when the latter course is adopted, this can well be in a two-way interaction with a similar attitude the other side of a recognized line of separation. The entrenched conservatives on each side will have a common interest.

Any recourse to violence will be liable to uglify such a situation, even if the immediate aftermath is back down not backlash. One has also to reckon with the frequently held though usually ill-founded belief that to goad the other side into being more extreme will be to render it unstable.

Not that Action–Reaction is some new discovery. Throughout History it has visibly figured time out of mind at every level from private contretemps to warlike crisis. Granted, an NEO cannot become less or more bloody minded except as perceived by those seeking to intercept it. But creating a requisite new world order is bound to involve plenty of such tension more generally.

An all too common illusion is that if a country or region is weak it will be less threatening. The USSR precipitated the Cuban missile crisis because she was becoming too weak awhile at the level of inter-continental war. See Chapter 7(1). The dismal aftermath of the aborted Arab Spring has left the Middle East more liable to jeopardise world peace than it would otherwise have been. And the threatened advent of biological warfare is liable to turn inside out the whole calculus of military strength. See Chapter 8(2).

Explosions

The destructive force of NEO impacts is expressed in terms of the nomenclature adopted originally to cope with nuclear warheads, either fission or fusion. The explosive energy (nuclear or chemical) and/or the kinetic energy (i.e. energy of motion) imparted by forceful contact are expressed in terms of the weight of TNT high explosive that would generate chemically an equivalent energy yield. See below. The kinetic fraction of any total will be a function of the square of the impact speed. Evidently, the extent of the material damage from a single impact or blast does not scale up anything like as rapidly as the energy thus released can. See Chapter 9.

What should also be borne in mind is that having an NEO break on impact into several menacing fragments takes far less energy than having it either melt or else be pulverized into bits so small as to pose no problems. Avoiding the large fragment outcome (e.g. by well-judged proximity fusing) will always be difficult.

Nuclear Reactions

The science of using nuclear energy for the destruction of life and resources was very fully related in an American official account first published in 1950.[2] Suffice now to outline certain key factors in the process. There are in existence close to a hundred elements, basic varieties of matter. All of them are comprised of atoms which consist of nuclei around which orbit electrons, each of these bearing what is conventionally termed a "negative" electrical charge. When an atom is in a state of electrical neutrality, the number of electrons orbiting a nucleus equals the number of "positive" charges in that nucleus. In a natural sample of a given element, there can be several forms – i.e. isotopes – that have identical chemical properties but are different in the character of their nuclei.

Any nucleus most essentially consists of protons (each carrying one positive electrical charge) and neutrons which bear no charge. The respective nuclei of two isotopes of the same element will have the same number of protons but a different number of neutrons. No two elements share similar isotopes.

The chain-reaction fission of certain isotopes high up the scale of atomic weights is associated with energy release offsetting some overall loss of mass. The same applies to the fusion of certain isotopes towards the bottom of this scale. However, the forcing together of proton particles required by fusion amounts to the joining of positive charges inherently countervalent one to another. This specifically requires energy output. It is supplied naturally within the fusion process we understand drives each and every star. But in our terrestrial ambience, a fission reaction is needed to get a fusion chain-reaction going explosively. The obverse side of this is that a fusion bomb can be made as large as its makers wish. It will not explode spontaneously.

May one note that a fusion reaction is often spoken of as "thermonuclear". And a bomb dependent on this process may colloquially be called a "hydrogen bomb". Peaceful applications of the nuclear process (whether fission or eventually fusion) require controlled moderation and so are harder to achieve *a priori.*

Biowar

A conflict mode the strategic community is altogether too prone to disregard is biological warfare. Reacting thus in the hope that this will cause it to go away is surely forlorn. Nor should we rule out the possibility that microbial life, benign or hostile, might be seeded through our galaxy on NEOs. Witness a rather rich mix of compounds we see as "organic" in a meteorite lately impacting on Australia.

Terrorism?

Some of us would feel more at ease appreciating the situation worldwide if the word "terror" and its derivatives were not now being used in a far more lopsided fashion than was once the case. Today any insurgent violence may be dubbed "terrorist" regardless of the circumstances, yet never any undertaken by a statehood regardless of the specifics. The term has thereby become like "racist", an outworn substitute for thought. This particularly applies when these words are used as nouns rather than adjectives.

The asymmetry in question is even less acceptable when applied in official and legislative documentation. If our language is to evolve in whatever direction, it should do so spontaneously. We can manage without steers from officialdom: American, Australian, British, Israeli or Russian, not to mention Syrian.

Simulation Studies

Still, the lead issue in this study has to be the physical threats posed to our Earth by comets, asteroids and meteorites.[3] Many simulation exercises have already been done, a primary aim being to enhance progressively our understanding of how reliably and precisely we are likely to assess particular collision prospects. Energy of motion is spoken of as "kinetic".

This work cannot usefully be pursued here in terms of specific instances. But a general question might usefully be raised more prominently. How acutely may the control of interceptions of NEOs be compromised by the distances and velocities liable to be involved operationally? A pulse of electromagnetic energy (light, wireless, radar, heat . . .) will take a third of a second to reach the vicinity of a NEO sixty thousand miles away, say. And in that time an NEO will move several miles.

This is a big complication for anti-NEO defence, nuclear or kinetic. At the December 2016 meeting of the American Geophysical Union, it was given as a major reason why interception would always be that much more difficult, given less than a year or so's early warning. For kinetic impact in particular, the velocity measurements and predictions would be especially time-consuming.[4]

Tectonics

Operationally speaking, the case is overwhelming for bringing tectonic events (i.e. earthquakes and volcanoes) on our Earth under the rubric of "cosmic" threats to be defended against. It gains pedagogic respectability in this context through their generic incidence not being confined to our mother Earth nor, in all probability, to our galaxy. Take Io, one of the four moons of Jupiter discovered telescopically by Galileo. It is 3630 kilometres across; and is covered in volcanic vents, something like ten of which will likely be active at any one time. Io's surface is continually being recoated thus.[5]

Stylistic Conventions

For purposes of enumeration, Anglo-American and metric units are employed interchangeably throughout this study. For that is how the supporting literature pans out. As implied above, prefixes indicating orders of magnitude (i.e. factors of ten) are used in the standard way. Thus "kilo" is the prefix for a thousand; "mega" for a million (10^6); "micro" usually for a millionth (10^{-6}); and "nano" for a billionth (10^{-9}).

Regions on Earth

In this text as in many others, the term "regionalism" may be employed in two opposite ways, the hope being that any ambiguity can be resolved contextually. The one meaning is the devolution of powers to distinctive areas within a particular statehood. The other, much used herein, is the institutionalized coming together of contiguous states within a given Earthly compass.

Brexit

As this study was being completed, the issue of inter-state regionalism came home to roost with a vengeance. It did so by way of Britain's deciding in the 2016 referendum to leave the European Union. The hustings had been less than satisfactory. Most of us learned far more about the issues involved in Brexit in the several weeks after the poll than we had known on the day itself. David Cameron was leading the "remain in" camp, yet could concede on television that the EU had already outgrown itself.

To many British voters, an overriding argument in favour of "exiting" was simply that the EU had waxed too large. But for many, too, an even stronger case was Britain's singular susceptibility to being swamped by immigration from within the EU as well as elsewhere. The fact that she had led the world in the Industrial Revolution (*c.* 1750–1850) allowed that (a) her population density was already high overall and (b) English had become the lead international language, one most intending migrants would start with at least a smattering of. However, the EU appeared solidly committed to the free movement of people within its borders, this to uphold the role of labour as a factor of production but also to preserve a basic liberal entitlement.

Nevertheless, this EU perspective may weaken over time because of how an on-going and marked rise in life expectancy together with automation will have different connotations for different EU members, this for reasons rooted in geography and history. Due balance may be preserved only if cross-border migration within the union is subject to control. My guess is that this will come within the next decade.

Among policy options subject to stringent review in the changing circumstances is one advocated further within these covers – namely, close Anglo-French collaboration in the sphere of thermonuclear strategic deterrence, using missile-firing submarines.

As regards submarines firing strategic missiles considered more generally, many of us have often argued that they can be a most stabilizing form of deterrence at long ranges. However, progress currently being made in underwater acoustics (alias sonar technology) may soon require some qualification of this understanding. Sonar is a sound sensing medium both sides can exploit. So can they sound suppression. So the balance of operational advantage may fluctuate over time. Even so, the overall conclusion needs likely be that, other things being equal, countries expecting to pay for a significant proportion of their deterrent outlays on submarines must be ready to spend still more on that dimension. See Chapters 6 and 8(2). One should further add the important aside that oceanic struggles for sonar supremacy could have a very negative effect on fishing and whaling.

One needs also to consider whether we might not be at the dawn of an era in which technological advance and resource availability will allow of strategic missiles regularly being used for non-nuclear strikes long-range, much as the Soviets threatened London and Paris during the Suez crisis of 1956. Naval anchorages could be especially vulnerable to this stratagem.

A consideration with more positive strategic implications is that a Britain still in Europe but not entirely of it could serve to preclude our continent's forging itself into a unity so solid as to become too assertive, situated as we are at the centre of the Earth's land hemisphere. The world needs regional confederacies that are not too close knit and tightly defined. How things are managed in a post-Brexit world will depend considerably on the United Kingdom's staying together and prepared to assume a varied and positive role within Europe and otherwise. She may, for instance, emphasise her role in the European Space Agency.

A relationship that should not be left to atrophy is the special governmental one between the USA and Britain. All else apart, it has well encompassed the notion that deterrence and arms control are opposite sides of the same coin. This duality became especially evident, in the period 1987 to 1997, in connection with defence against strategic ballistic missiles. It was a subject of considerable debate between Washington and London, with the latter regularly showing more concern than the former about avoiding the weaponization of Space via the disposition in orbit of downwards-directed laser interceptor stations.[6] It is to be observed that, despite a growing sense of a need for regional anti-ballistic missile defence by means of surface based weapons, neither the USA nor any other nation state is overtly making ready to place any such engagement facility in orbit. So long as it lasts, this state of affairs augurs well for commonality in anti-NEO defence.

But how long will it last? Received opinion, in the Anglo-American media at least, has lately been inferring, either gleefully or ruefully, that further national militarization of Near Space is already on its way. Undeniably, governments need to be giving management of the Near Space milieu much more urgent attention. But not in the direction of unilateral anti-terrestrial rearmament. Rather it should be in pursuit of multilateral NEO crisis precautions which also support wider planetary aims.

The Trilogy

The books that make up the trilogy Sussex has lately published under my name bear on this analysis continually. Duly, the texts in question are recorded in the annotations via the initials of their titles. So for *EC* read *Engaging the Cosmos*. For *GHC* read *The Geography of Human Conflict*. For *BL* read *The Bounds of Liberalism*. Page references are attached in each case. The three volumes came out in 2006, 2009 and 2014 respectively. The aim has been to have them achieve together an overarching perspective.

Appreciations

Siân Prosser, the Librarian and Archivist of the Royal Astronomical Society, has been invaluably supportive throughout. Also very helpful have been the library staff at the International Institute for Strategic Studies and the British Library.

Jill Wells, my literary secretary of 44 years standing, has lent yet again unswerving and most insightful support, backed solidly by her husband, Trevor. My wife, Yu-Ying, has likewise been a pillar of strength, notwithstanding some exceptionally challenging commitments as the retired head of the invaluable Japanese collection in the British Library.

This study we have dedicated to the late General Sir John Mogg and his widow, Lady Margaret, our respective families having long been close friends in the townlet which is home to both sides: Watlington in Oxfordshire on the spring line of the Chiltern Hills.

Each extended family has had its share of world war. But come the time, John and Margaret were able through NATO to play a signal part in bringing the Cold War firmly to a conclusion. Then, entered upon retirement, they were to prove zealous in helping to preserve in

Watlington in changing times a community spirit based on expansive inclusivity.

The Author

In 1994 Professor Neville Brown became a senior member of Mansfield College, Oxford. His career has been based on the interaction between the humanities and physics, especially the sky sciences. Coming in from physical science, he read economics with geography at University College London (UCL) followed by modern history at New College, Oxford. For half of the time he then spent as a forecasting officer in the meteorological branch of the Fleet Air Arm (1957–60), he specialized in upper air analysis. Other assignments included a British coastal air station plus gunnery trials and junior staff duties with the Mediterranean Fleet. He was a field meteorologist on two British Schools expeditions to sub-polar regions. His interest in military strategy has its roots in a wartime childhood but was made professional by naval service and a subsequent lectureship (1960–2) at the Royal Military Academy, Sandhurst.

In 1980, he was elected to a chair in International Security Affairs at the University of Birmingham. He has held Visiting Fellowships, or the equivalent, at the UK National Defence College, then at Latimer; the School of Physics and Astrophysics at the University of Leicester; the International Institute for Strategic Studies in London; the Stockholm International Peace Research Institute; the Australian National University, Canberra; and the Defence Engineering Group, UCL. From 1965 to 1972, he worked part-time but quite pro-actively as a defence correspondent, focussing mainly on the Middle East and South-East Asia. He was accredited to the *New Statesman* (1965–8) and the *New Scientist* (1968–70). He also filed regularly for *Le Monde Diplomatique* and the *New Middle East*.

From 1981 to 1986, Neville was the first Chairman of the Council for Arms Control, a British all-party body drawn from parliament, the churches and other professions and committed to a serious but essentially multilateral approach to arms control. He became then deeply involved in the multinational debate about Ballistic Missile Defence (BMD).

From April 1994 to the summer of 1997, he worked half-time in the

Directorate of Sensors and Electronic Systems (within the Procurement Executive, UK Ministry of Defence) as the Academic Consultant to the official Pre-Feasibility Study on what policy Britain should have on strategic BMD. A declassified version of an 87,000-word *Fundamental Issues Study* he wrote during this assignment was published by Mansfield College in 1998. Throughout, the vexed question of Space-based missile defence was on the agenda. He has remained firmly opposed to Space weaponization for whatever purpose.

In 2000, Neville was the editor for a very seminal Sino-European Conference in Beijing on BMD. The thirty or so officials and academics present were senior people with directly relevant expertise. The European contingent came from France, Germany and Britain.[1] About half the papers presented and published together in English were by our Chinese colleagues.

He has authored or edited some twenty books or major reports. Among them has been *The Future of Air Power* (Beckenham, Croom Helm, 1986). But with the award-winning *Future Global Challenge* (New York: Crane Russak, 1977), he began to give economic, cultural and ecological factors more salience in the quest for a peaceable world. The recent trilogy of studies carried this progression quite a bit further. See the Prefatory Statement. His chief contribution to date to the scholarly study of climatology has been *History and Climate Change, a Eurocentric Perspective* (London: Taylor and Francis, 2001).

In 1990, Professor Brown was elected a Fellow of the Royal Astronomical Society. In 1995, the University of Birmingham conferred on him a Doctorate of Science in Applied Geophysics.

1

Legends and Realities

Our solar system is the spread of heavenly bodies (planets, moons, meteors, asteroids and comets) travelling round our parent star, the Sun. The indications are now strong that very many stars within our universe, our cosmos, have such a panoply.

An assemblage of billions of stars within the cosmos is a galaxy. Clusters of galaxies are also discerned. So are "superclusters" – i.e. clusters of clusters. The Milky Way is our galaxy; and has some 400 billion stars. Our universe has several billion galaxies. And there are now generally thought to be many universes, quite possibly an infinity.

The lead question addressed herein is how we might best make ready to cope with what were identified above as Near Earth Objects (NEOs): that is to say, planets, asteroids and comets – especially those which have been diverted from their orbital tracks by collision and/or gravitational attraction (e.g. by the great planet Jupiter) and are deemed to be heading too close to ourselves for peace of mind. Strong meteor showers are a plurality also within this remit.

Interception, hopefully culminating in virtual destruction or decisive deflection, should be the "front line" or "active defence". But if, as is the case, studies on this score to date by no means consistently reassure, we should prepare ourselves for a high level of "passive defence", minimizing the damage and life loss as and when NEO impacts actually occur.

An added attraction of "passive defence" preparations is that they may contingently be directed against disasters terrestrial in origin. One variously thinks of supervolcanoes, earthquakes, tsunami, magnetic reversals, hurricanes, rampant epidemics, plant disease, and nuclear war. The frequency of NEO threat presentation seems set to vary randomly during the two centuries ahead, which is about the time frame we should actively concern ourselves with. At present, no major menace appears to be in the offing within it. But we cannot be quite sure. What, for instance, about asteroid Bennu? See Chapter 12.

Science Fiction

The ancient and widespread tradition that comets are harbingers, probably of doom, owes its all to the way their tails turn luminous as they draw relatively close to the Sun. In 1705 Edmund Halley, the British astronomer, accurately predicted 1758 would be the return year of the comet which now bears his name. But that scientific aplomb did little to weaken doom-laden portent. Before long, the idiosyncratic unfolding of science fiction would reinforce it a good while.

In *The Comet* (1832), the American physicist and poet, Oliver Wendell Holmes, envisaged utter destruction all round, hard upon failed attempts to disintegrate by gunfire a comet closing in. Among other renderings around that time was one downright fantastical by Edgar Allan Poe, *The Conversation of Eiros and Charmion* (1839). He has a comet passing near Earth suck virtually all the nitrogen from the air, leaving the then almost pure atmospheric oxygen to combust vigorously.

H. G. Wells was to write along similarly ludicrous lines in 1897 and 1906. In the latter text, *Days of the Comet,* gas from the NEO induces euphoria in us Earthlings and a consequent rise in altruism. Yet soon he was to have a flash of uncanny prescience when in *The World Set Free,* published in 1913, he postulated the outbreak of atomic warfare soon after the requisite technical breakthrough in 1953: "Power after power . . . went to war in a delirium of panic, in order to use their bombs first. . . . By the Spring of 1959 from nearly two hundred centres . . . roared the unquenchable crimson conflagration of the atomic bombs". Even so, Wells' interest in military science as such was too lay and too episodic to yield any very definitive results. Instead he oscillated between a blimpish insensitivity to change and a skittish indifference to continuity.

His successive attitudes to "air power" (a term he appears to have coined) bear out this judgement only too well. Let us look first at two books he published early in the last century, the decade which saw the historic flights of the Wright brothers (1903) and Louis Blériot (1909). Both in *Anticipations* (1901) and in *The War in the Air* (1908), he does accord this domain of warfare an overriding importance that broadly corresponds with our subsequent experience. However, he foresees the exercise of air power as being through echelons of fighting balloons "as thick as starlings in October".

Although in *The Shape of Things to Come* (1938) Wells visualises the massive use of aeroplanes in counter-city bombardment, his preliminary struggles for supremacy between rival air fleets tend to be too drawn out. Thus he has the German air force take over a year, in the early 1940s, to overcome the Polish.[1,2] Aerospace prediction is now far more systemic. But wide margins of uncertainty remain, not least in the NEO ambience.

In 1897, George Griffith foresaw (in the *Great Crellin*) a giant gun being successfully designed and deployed to divert a comet. But overall it cannot be said that over the years NEO interception has had much fictional prominence. Griffith thus provided the only instance cited by Mike Ashley in the latest of his enriching reviews of science fiction.[3] But lately the said genre has moved up the lists, encouraged by Arthur C. Clarke (1917–2008).[4]

Pluto and Mars

Pluto, the quasi-planet within our solar system, is agreeable to cite at this juncture in that it rates as a most solid achievement by Percival Lowell (1855–1916). He was one of two American astronomers who predicted the existence of such a heavenly body beyond Neptune; and the laboratory he founded in Flagstaff, Arizona, was whence it was first observed in 1930. But, alas, the prime justification for mentioning Lowell at this juncture has to be that he also afforded our most graphic instance to date of cosmic hype.

In 1877, an Italian astronomer, Giovanni Schiaparelli, had claimed he had discerned a criss-cross spread of *canali* ("channels", meaning water courses) on Mars. Then, in 1906, Lowell capped some years of fanciful speculation about this supposed find with his book, *Mars and its Canals*. He proposed that dwellers on the "red planet" had dug these features to carry water for irrigation from its polar ice caps. Yet canals of reasonable width would have been much too narrow to be overtly observed by any telescopes then installed on Earth. Nor would they have been discerned at all easily if set within broad avenues of trees. Nor was any convincing cartography presented. Yet despite endless protestations by Lowell's more responsible colleagues, the popular influence of this hypothesis was considerable, certainly across the Anglophone world, through 1940. And, in a shadowy way, it lives on yet.

Velikovsky and von Daniken

Nor was this to be by any means the most way-out example in this field of surreal inventiveness.[5] In 1950, Immanuel Velikovsky, an American psychiatrist, published his *Worlds in Collision*. The ructions this book caused among the astrally inclined had by 1970 fed into a prolonged dispute, sustained by the Vietnam war, over the whole ambit of "big science".

The Velikovsky imaginings told how, before entering its present orbit *c.* 1500 BC, the planet Venus was a giant comet that had somehow ushered forth from Jupiter. Twice in 52 years it had passed close to the Earth, the first occasion being during the Jewish exodus from Egypt. Supposedly, too, the Earth's oilfields were "partly" deposited by the cometary tail.[6] Among other depositions – said Velikovsky – were locusts, vermin, hail, manna and meteorites.

How then could anyone possibly imagine a comet so huge and enriched as to transmute into the Venusian mass; and be able, at the same time, to build up oilfields? The author tells in the preface how this vision had come upon him in 1940. That much was believable.

Then there were the best-sellers by Erich von Daniken starting with *Chariot of the Gods?* His contention was that, between 40,000 BP and 592 BC, the Earth was visited by a majestic breed of astronauts. A fierce critique of this fantasy ends, none the less, with food for serious thought: "If scientists can learn to communicate their discoveries with the same enthusiasm as von Daniken; and if we can learn to partake of the tree of knowledge without losing the innocence and curiosity that prompted our first questions, the human spirit will evolve".[7]

An aberrant contribution came unexpectedly in 1974 from John Gribbin with Stephen Plagemann. Their advice then was that "A remarkable chain of evidence . . . points to 1982 as the year in which the Los Angeles region of the San Andreas fault will be subjected to the most severe earthquake known in the populated regions of the Earth in this century".[8]

The nub of the argument, as presented, was that, in the said year, an almost exact alignment of all the planets then on the further side of the Sun would trigger a tidal surge of molten rock within the Earth's interior. Though this inference was not that counter-intuitive, it had to be challenged on several counts. But the crunch criticism could not be just that this direct consequence failed to materialize. It was and is that it never could have. Tides of water, air or molten rock do not simply depend on the overall gravitational pull exerted on the Earth by another heavenly body. Instead, they derive from the differences in the strength of that pull on points on or in the Earth short distances apart. A small body fairly close by may well exert more influence tidally than a large one further out. The Moon determines our ocean tides much more than the Sun does.

Forward Thinking

Obviously, the bulk of actual astronomical research and rapportage is far more sound than such digressive exemplars might suggest. Suffice now to note that John Gribbin has emerged as an invaluable communicator of astral science to lay people. So how will the lay debate shape up, as the anti-NEO defence issue assumes more salience? China versus the rest? America versus the rest? China and America versus the rest? Big science versus customary insights? Activism versus denial? Planetary versus multinational?

Will it necessarily be seen as especially liberal to favour heavy expenditure on an anti-NEO screen? What will be seen as Leftist, assuming the Left–Right distinction continues to be accepted in other regards? How much will turn on the specifics of any operational posture: surveillance and interception; command and control. . . . There will be much scope in the decades ahead for interdisciplinary fiction sound enough to point towards helpful prediction.

Impotent and Obsolete?

In March 1983, President Ronald Reagan launched the Strategic Defense Initiative (SDI), the programme for making ready to destroy in flight strategic ballistic missiles, Earth-launched. His declared aim was to render nuclear weapons "impotent and obsolete". Yet quite apart from the accepted fact that a clear majority of the several thousand nuclear warheads already in service within the Atlantic Alliance alone were for local war use (nuclear mines included), this was mission impossible. Nobody this author ever met in Washington, London or anywhere else, in the course of a four-year governmental study (see Prefatory Statement), thought one could never expect anything close to a 100 per cent rate of missile interception at the top strategic level. A decisive reason proffered was that you would never know definitively what pattern of attack Moscow or whoever was adopting, given what would likely be a novel and fast-breaking crisis.

Approximate Analogy

An active asteroid or comet threat would almost certainly be posed by a single incoming object. But one might not have much chance to gauge its precise character. What are its dimensions? What is its composition? What is its coloration? How firmly might it hold together if impacted? How many interceptor missiles might you have to dispatch to stand, say, a 90 per cent chance of impact? And if interception should involve either skewed deflection or random disintegration, might not the outcome be to make matters worse?

Above all, how much might one guestimate with assurance? And, indeed, how could you know if your confidence was well founded? It would never be easy to respond with a satisfactory blend of knowledge and surmise. Hence a need for passive defence, starting with visual or radar early warning and leading on to evacuation and disaster relief.

Initial Ruminations

Visiting Washington in 1995–6, I was told some thought had already been given to NEO interception. This postscript was duly added to my Ballistic Missile Defence (BMD) report to the Ministry of Defence in London:

> "There is one further issue to consider, one outside the defined remit of this *Fundamental Issues Study*. Should one be working towards a system which might perhaps be called GPACE (Global Protection Against Collisions with Extra-Terrestrial objects – alias meteors, asteroids or comets). This subject is a function of science fiction far less than the Unidentified Flying Objects (UFOs) or 'flying saucers' various Ministries of Defence have kept a watching brief on. Of late, indeed, there has been a confluence of evidence from astronomy, archaeology, history and historical climatology that real collisions will have compromised life on Earth much more often and more severely than used to be realised.
>
> This is not something to pursue further at this juncture. The technological parameters are too different from those of any recognized mode of BMD. Besides, one would have to judge stringently the timing and political setting of any diplomatic initiative."[9] Attention was drawn then to China's 1996 intimation that she might make the "asteroid threat part of an alibi for not signing a comprehensive nuclear test ban."[10]

Substitute for War?

However, there is a further aspect to weigh. Writing in 1902, hard upon that turn of century when the whole of humankind seemed to be merging into one congested though vibrant global entity, see Chapter 4(2), William James (the theological brother of Henry, the novelist) addressed the need to find a substitute for war. He acknowledged the quiet pride to be derived from heroism. Nevertheless, he felt constrained to insist that we "gravely" ask whether "this wholesale organization of irrationality and crime can be our only bulwark against *effeminacy*" (my tremulous italics).

Instead, we were advised that what we needed to "discover in the social realm is the moral equivalent of war: something heroic that will speak to man as universally as war does and yet will be as compatible with their spiritual values as war has proved itself to be incompatible. . . . May not voluntarily accepted poverty be 'the strenuous life' without the need of crushing weaker peoples?"[11]

A number of reasons might be adduced for responding in the negative. Thus soldierly honour does not consist simply or even preponderantly of being prepared to kill the other lot. It must also involve a continual preparedness to put oneself in harm's way. To do so not immoderately can be quite exhilarating.

The fulfilment paradox thus presented is highlighted by the not uncommon concern of gifted academics to savour the "hands on" excitement of modern war along with that of scholarly creation. A striking case in point was, as a young philosopher, Ludwig Wittgenstein (1889–1951). When World War One broke out, he was starting work on his *Logical-Philosophical Treatise*, the only book he was to get published during his lifetime. He volunteered for the Austrian army forthwith, a problematic medical assessment notwithstanding. He was to be continually on active service, virtually until the war's end. His specialism throughout was that of an artillery observer in No Man's Land, a well regarded role that brings out firm but quiet courage. He was awarded a Silver Medal for Valour.[12]

Acceptance of comparative penury for the benefit of the more deprived is not a straight alternative. Granted, such a disposition can well up in the aftermath of war. But when it does, it will be to welcome the conclusion of hostilities, not a belated alternative to their having been waged in the first place. As to whether charity might reduce the risk of future warfare would depend, case by case, on a variety of specifics.

Planetary versus International

Prominent among the wider attractions of what might be judged a sound NEO screen could be that it might place the term "planetary" near the centre of our thought patterns and discourses. In 1972, Zbigniew Brzezinski (to be National Security Adviser to President

Carter, 1977–81) wrote of the contrast between the time-honoured "international approach" to world affairs and the more "planetary" perspective favoured by radicalized youth. He proposed that political leaders guided by new concepts were needed to bring the two together.[13]

Brzezinski came across then as perhaps too aware of the more futile and anarchic aspects of the "planetary" approach alone, maybe on account of his Polish-American concern with the containment of Soviet influence. But may one just say that his call for a new synthesis had encouraging operational precedents. Thus after World War Two, George Marshall and Douglas MacArthur became visionary peace-builders in Europe and Japan respectively. During America's war years, they had been great military strategists.

There may be room, too, for a deeper philosophical perspective, particularly if the role of an anti-NEO screen is not interpreted too prosaically. At the height of the SDI controversy, Patricia M. Mishe, the founder of Global Education Associates, argued that President Reagan's Strategic Defense Initiative was a morbid diversion from the proper study of Space, a study that otherwise might help us understand better our human situation. She concluded her *Star Wars and the State of Our Souls* as follows: "We are the soul of the Universe. How each of us responds to the conflict in our souls is what we will bequeath to the universe, that is what will become the soul of the universe".[14]

2

Galactic Reconnaissance

The legend that Mars awaits, almost invites, mass colonization by ourselves rumbles on yet. And if not Mars, then somewhere else within our solar system. And if not within our solar system, then on landfalls far beyond, whenever.

In 1977, a ranking Princeton physics professor, Gerard K. O'Neill, influentially foresaw, "perhaps by the middle years of the next century", mass emigration to Space. Within the first decade, 290,000 might be settled; and after the third, the total (with natural increase on location) might be 631 million.[1] In 2016, a possibly habitable planet was discovered orbiting *proxima centauri*, the star nearest to ourselves, after the Sun.

However, those O'Neill findings were altogether too sanguine and exact. But it may still be instructive to ask what the trade-offs might have been. Very heavy capital outlays would have been incurred in the getting of every migrant settled and in gainful employment. No less worrisome would have been the loss to their parent Earth of men and women imbued in good measure with the game-for-anything spirit of Yuri Gagarin and Neil Armstrong.

Two years after this O'Neill projection, Freeman Dyson, another prominent and eccentrically engaging doyen of Princeton astrophysics, proposed that a substantial and variegated human exodus would be a good way to preserve the cultural diversity continually being eroded across the Earth itself.[2] Yet whether it could ever have fulfilled this purpose well, or indeed at all, is moot. Again, too, one needs reckon with the diversion of material resources and of exceptional human calibre.

Besides which, travel in any long-range Spaceship would require the continual imposition of a tight behavioural uniformity, not the cultivation of cultural diversity. Moreover, any time spent ashore on some inclement heavenly body would accentuate this need. So more spartan than Sparta was or ever could have been? In that same study, Dyson found O'Neill's logistical reasoning altogether too sanguine.[3]

Meanwhile, Britain's Astronomer Royal has seen colonizing the heavens as the best post-2100 prospect for the survival of our species in the face of genetically-engineered epidemics spread by maybe just "one fanatic or weirdo with the mindset of those who can design computer viruses".[4] But how could we be sure the guilty party or its pals had not infiltrated a mass emigration flow? And, in any case, would not a counter-measure on that scale feed dangerously the vanity of radicalized malcontents?

Space and Time

All the same, the biggest obstacles to Space travel beyond our solar system are not the above contradictions. They are sheer Distance and Time. According to Albert Einstein's theories of Relativity, which still hold up well, the maximum velocity matter could ever travel at is 186,200 miles per second. This is also the speed at which electromagnetic energy (i.e. the heat, light and radio spectrum) radiates *in vacuo*. A "light year" is the distance this "velocity of light" covers in the course of an Earth year. *Proxima centauri* is 4.28 light years away. A round trip to its vicinity and back, not at the velocity of light but perhaps in due course at 20 per cent of it, could take several decades.

However, a further Einsteinian inference was that, as the speed of a moving object approaches more closely that of light, Time itself is stretched progressively more within it. A voyager on board a future Space rocket might finish appreciably younger than they could otherwise have been. But this might not become observable except at relative speeds not yet attainable. Proposals to feed energy for interstellar propulsion by means of ultra-powerful terrestrial laser beams may discount unduly the tendency of all electromagnetic beams to diverge, however short their wavelengths. In any case, the technology required is still far ahead of what is currently operative.

Not that propulsion would be the only radical challenge presented by interstellar travel. Navigation would be another. So would any attempt to land on a planet or moon; and, perhaps tougher still, take off smoothly again. There would also be a contradiction between the quietude required of a crew during those long interstellar years; and the energy and daring likely called for at destination or when coping with outboard mission faults throughout. The psychologists and anthropologists must advise us. But it looks *a priori* as if two contrasting personality patterns might be required within the same persona.

The likelihood of alien life being encountered during a particular landing may be small. But if any with intelligence were found *in situ*, it could well turn decidedly hostile. Alien microbes might prove exceptionally threatening to Earthling Spacefarers and hence to Earth life in general. The same might apply with our microbes *vis-à-vis* indigenous aliens. May one add that human beings ill protected by atmosphere, Space suits or structures will be considerably at risk from debris or from streams of charged atomic particles travelling through Space at perhaps 50,000 miles a second. Space suits need to be a dozen or so special layers thick. And, when fully exposed, might still be inadequate versus galactic particle radiation.

Telescopic Revolution

More extensive Space reconnaissance (unmanned or manned) will have a valuable anti-NEO role to play. Already some 3000 asteroids regarded as potentially menacing have been logged.[5] But will it not be as crucial to a fuller understanding of dark energy, dark matter, the nature of black holes, the number and basic characteristics of moons elsewhere, the early history of our universe and so on? That knowledge will considerably be furnished near term by (a) a veritably worldwide and lunar distribution of large radio telescopes and (b) the advent of more telescopes mounted in satellites orbiting virtually free from terrestrial light or infra-red interference.[6] As regards the satellite mode, the Hubble Space telescope has been an excellent pioneer programme. Its near-circular orbit has been *c.* 560 kilometres above the Earth's surface.

The European Space Agency's Gaia dual Space telescope, launched in 2013, aims to measure the distances, parallaxes and motions of some of the many billions of stars in our Milky Way galaxy and throughout the rest of our Local Group of some thirty galaxies. Data flow began in 2016.[7] Henceforward the construction costs involved in such programmes are expected to be reduced substantially by making modules which can be inflated in orbit.

The International Space Station is a programme initiated by the United States. The European Space Agency, Japan and Canada were successfully invited to participate in 1984, albeit against a background of widespread scepticism as to the project's operational utility. Russia joined after 1990. The sceptics are still exercised.

The station currently has 15 modules, its total mass being 420 tons. Its construction to date has cost roughly 150 billion dollars. Within a decade, it is expected to receive another five modules, these being inflatable after attachment.

Also in train is more reliance on the principle of coordinating on Earth arrays of telescopes within a very limited area and/or much more widely. The Cherenkov Telescope Array (CTA) currently being developed will observe particularly the high energy spectra of the universe. It has become already a "truly global effort by well over 1000 scientists from 29 nations".[8] Its early aims are to elucidate the origins of cosmic radiation; particle acceleration around black holes; and the fundamental nature of matter.

No Mass Migration

We would do well simply to accept the outer limits of the solar system as constituting the boundaries of our galactic home, limits we expect never much to exceed in terms of a substantial human presence. That could accord well enough with the objective considerations identified above. It could also hark back to our primeval propensity to live within defined tribal areas. But such acceptance still more firmly excludes mass galactic emigration being seen as a remedy for the difficulties we have created for ourselves through mismanagement of the biosphere of our parent planet.

No other planet or moon accessible within the solar system seems likely, in any case, ever to accommodate more than a tiny fraction of Mother Earth's present human population, most likely nothing more than teams of specialists managed and victualled from the Earth itself. Our Moon has no atmosphere at all, nor any water except perhaps for what may be some very limited cometary depositions. Venus was thought by some to be a possible home from home until, in 1932, it was finally established that its atmosphere was almost entirely carbon dioxide, this producing massively a "greenhouse effect". Mars still has water or, at any rate, brine. However, its atmospheric pressure is but a few per cent of ours. And it is subject to dust storms, which may be planet-wide, lasting for weeks.

Giant Planets

Manned visits to certain moons of the giant outer planets, Jupiter and Saturn, have been considered since early in the Sputnik era. And the relative abundance of elements used in life which are found in certain locations out there has been noted. By 1958, Freeman Dyson had developed a special interest in Enceladus, the next-to-innermost moon of Saturn. It is only 500 kilometres in diameter. Moreover, its parent planet would look *c.* 3000 times as big to an observer on Enceladus as our Moon at the full does to ourselves. Still, this Saturnian moon might at least be a viable source of water and certain other substances.[9]

However, salient features of Saturn and Jupiter indicate that neither would be welcoming to ourselves. On Saturn, gales can blow at up to 1800 kilometres an hour. The mean temperature on Saturn's surface is 134 Kelvin (i.e. 140 degrees of frost, Celsius), while that on Jupiter's cloud tops is *c.* 130 K. The escape velocity is 36 kilometres per second from Saturn and 60 km per second from Jupiter as against a mere 11 from the Earth. May one add that hardly any of the moons orbiting a solar planet seem to have an atmosphere.

To my mind, these successive comparisons cap the argument that even within the solar system, we can only sensibly envisage brief visits (regular or occasional) by specialist teams, there never being colonies able long to sustain a substantial measure of workaday autonomy. What this will come to mean in terms of the number of people earmarked to serve beyond the Earth at any given time will depend chiefly on what military posture is adopted overall for anti-NEO defence. But ten thousand might be as good a stab as any for two centuries ahead. The operational and political case for having short stay rosters would be strong. Three month tours the norm?

Artificial Platforms

Cosmic Command military requirements might well be the prime determinants of the extent to which man-made modular platforms are deployed in Space over time. But civil purposes will need to be served as well: meaning, in particular, facilities for easing the global warming problem, either by focussing solar energy on doing terrestrial work that might otherwise be done by fossil fuels or else by simply reflecting back outwards solar energy incoming. Large telescopes on Earth or in orbit might come to have both a research and a Cosmic Command role. So ultimate authority to rest with the latter?

Correcting for Greenhouse

An array of solar cells locked together in a high Earth orbit may absorb virtually 100 per cent of the solar energy coming directly towards it. This is on account of there being little or no interference from clouds or atmospheric gases. The solar energy incident in such an ambience could be 32,600 watt hours per square metre per day, whereas the inflow therefrom directly onto the Earth's surface has been given as 8,600 in sunny Phoenix, Arizona; and 3,500 in damp and cloudy Boston.[10]

However, there seem to be no plans extant currently for proceeding down this route. Instead, doubts have been expressed on several counts. Any microwave transmission to the ground of energy gathered thus in orbit would probably not exceed 60 per cent efficiency. Among other objections are said to be an orbital mass of several tons needed for every megawatt to be generated; heavy assembly costs; and the possible effect of microwave overspills on radio transmissions. A good size oil or coal-burning power station on the Earth's surface might generate 2000 megawatts.

A raft of technical objections has likewise been raised to the alternative concept of limiting greenhouse warming by the reflection of a proportion of the Sun's rays back into Space by means of a diaphanous mirror parked in low Earth orbit. However, as regards whichever approach, an overriding constraint, as of the present time, continues to be a failure to create the right international political context. Nevertheless, in the United States and Britain at least, a debate on the operational aspects has become more insistent this decade. Witness a recent IISS overview.[11]

Mining and Manufacturing

Much the same applies to such other anticipated uses of Near Space as the mining of asteroids, platinum being a scarce metal regularly mentioned. Donald Yeomans has done an especially valuable assessment of the outlook in this regard, presuming the while that "the planet Mars and its two moons are the ultimate destinations for human exploration beyond the Moon".[12] Landing an unmanned probe on a comet has already proved possible.

Recovering minerals thereby should generally become more attractive, or less unattractive, over time. But uncertainties remain. Thus the seabed in marine Economic Exclusion zones may be a terrestrial alternative source of fresh supply.

A third of a century ago, the possibility of micro-manufacturing in Space was ardently foreseen.[13] The virtual absence in orbit of gravity appeared to leave more scope for quality control. Moscow was especially sanguine on this score.[14] For some years now, however, little has been heard in this regard, at any rate in the West. One reason may be a fuller recognition of the physiological benefit human operators may derive from at least some rotational gravity.

Yet this is not to deny that the said theme may lately have been staging something of a comeback. It could be so doing by way of conflation between mounting interest in private Space manufacturing and in Space tourism. Unfortunately, in the literature there is almost certainly a tendency to exaggerate the extent to which Space colonization might be supported by the range of raw materials available out there.

Tourism in Space

Giving tourists even the shortest of partial orbital flights (at a per capita cost of something like $150,000 a time) would be to flout everything now being said about conserving resources so as to undermine the global ecology and economy as little as possible. Two of the visitors thus far accommodated on the Space station awhile are said to have been paid for with tens of millions of dollars.[15]

Cosmic tourism also flies in the face of a concern to limit the widening of income differentials now under way very generally, within countries and between them. Sir Richard Branson has long enjoyed a well deserved repute as one of Britain's most progressive entrepreneurs. But it is hard to feel good about his roseate vision of the advent of Space tourism as "wonderful news for the millions of individuals of different abilities, ages and nationalities who dream of seeing their home planet from the blackness and silence of Space". One likewise feels concern about his collateral claim that "Before long we expect to be able to launch as many as 100 small satellites in a 24-hour period".[16] The second declamation is not linked tightly to the first but is likewise too effusive.

The Military

On the military side, there are primary distinctions to draw. One is between the weaponization of Near Space and leaving it free of actual weaponry, though not of civil or other facilities also usable by the military for surveillance and communications.

Another is between any weapons present having been respectively designed to intercept either NEOs or else terrestrially-based missile systems. Here the prime consideration is that the former will very generally have to have much longer detection and engagement ranges. The nudgings aside of asteroids may have to begin several orbital revolutions ahead of their predicted collision with the Earth. More urgently, if an incoming NEO is liable to fragment on interception, enough time must be left for menacing fragments to disperse sufficiently before the Earth's vicinity is reached.

To which one might add that certain types of warhead might be unsuitable against NEOs travelling at exceptionally high speeds through what will largely be an extremely thin outer atmosphere. This could readily be the case with proximity-fused ordnance, nuclear charges included. Blast effects will be much too weak; and times in contact far too short. See Chapter 9.

Military Reconnaissance

The 1962 Cuban missile crisis has been cited in the Prefatory Statement as having had its immediate origins in the recognition on each side of an imbalance, at the intercontinental level, between the thermonuclear missilries of the two Superpowers – the Americans then being sure, for some years ahead, of a decided advantage. The situation was one which for a while could have induced either side to launch suddenly all-out war.

Quite often, however, definitive overviews will engender mutual reassurance as between adversaries. A case in point may be the downward revision from 1956 of the West's estimate of the Soviet army's order of battle. Received opinion soon brought it down to 150 divisions after mobilization, instead of the 450 then being cited by no less informed and temperate an authority than Sir Basil Liddell-Hart.[17] The said revision was facilitated by the illicit Lockheed U-2 manned aircraft overflights of the USSR from 1956; and confirmed, from 1961, by licit reconnaissance from Space.

Space Debris

Still, much may ultimately hinge on our peacefully negotiating our way out of the problem of orbital debris, especially in the altitude band of 400 to 1000 kilometres, the International Space Station being at 450 kilometres. Well over 6000 satellites have been launched since 1957, many of them no longer operational yet still orbital.

Around 20,000 pieces of debris ten centimetres or more across remain in orbit as do well over 100,000 smaller but still troublesome bits. In January 2007, China saw fit to conduct an experimental anti-satellite interception, thereby breaching a moratorium in that regard informally maintained since 1982.[18] This single action added 2300 fragments of golf ball size or above. At least a third of these will continue to circle the Earth through 2035. More work needs to be done generally on driving debris Earthwards by means of deceleration sweeps.

Space Management

Lately, there have been a raft of reminders that the military management of Space is bound to become either rather more collaborative or else a lot more tension-ridden, not to say anarchic. NATO's Supreme Allied Commander, General Denis Mercier, describes Space as "a domain that is going to be as important in modern warfare as any other. We have a responsibility and a necessity to work on it".[19] Suffice to recall that, since the 1960s, NATO has been a leader in treating deterrence and arms control as the two sides of the same coin. The intricate if ethereal nature of this duality as applied in Near Space can be no justification for disregarding it there. Yet one fears disregard may now be taking over.

We need urgently to move beyond the UN Outer Space Treaty launched by the USA, the USSR and Britain in 1967; and currently subscribed to by over 100 states. Its inspiration was the Antarctic Treaty of 1959.[20] It lays down certain cardinal principles for the use of Outer Space, including the Moon and other nearby celestial bodies. Weapons of mass destruction are currently excluded. So, too, are national claims to extended spatial zones. Forms of collaboration include information exchanges.[21] The first and last precepts should now be regarded as something to build back or up from, a celestial "reconnaissance in force".

Drafting the Outer Space Treaty has throughout been rated the most important achievement of the Committee on the Peaceful Uses of Outer Space (COPUOS) set up by the UN in 1958 (with 18 members initially), this against the background of the International Geophysical Year (IGY). Other products of COPUOS in those early Space Age years included the 1968 Agreement on the Rescue of Astronauts; the 1972 Convention on the International Liability for Damage caused by Space Objects; and the 1976 Convention on the Registration of Objects launched into Outer Space.

To start with, there was some dispute about the upward limit of national air space. But this was soon resolved, informally though sufficiently, with the USSR preferring to set the horizontal divide at 100 or 110 kilometres and the USA at 110 or 112.[22] Either way, much human activity above the divide (single nation or multinational) directly affects our everyday world below.

A primary physical boundary is the "tropopause" – the sharp natural break between the troposphere (i.e. the lower atmosphere) and the stratosphere. Being almost free of water droplets or vapour, the stratosphere does not wash down solid particles that have entered it. These therefore tend to descend very slowly, the smaller often taking five to seven years to pass through to the troposphere below. The tropopause is averagely at about 15 kilometres above sea level near the equator but below ten kilometres towards the North Pole.

3

Tough Centuries Ahead

The next two centuries or so could well prove critical to the outlook for we Earthlings over a much longer term. Even without impacts from Near Earth Object (NEOs) or comparable tectonic events, this prospect could still be in train, given the unprecedented array of economic, social, technological and philosophical changes under way and gathering pace. But a sense of looming crisis is reinforced by a greater awareness than hitherto of the cosmic and tectonic disturbances known to have come upon us, sometimes with little or no warning. It is surmised, for instance, that a "billion-ton" comet probably passed within several hundred kilometres of the Earth as recently as 1883.[1]

Can we humans, as the lords of Spaceship Earth, gain the capability to respond to such threats as may materialize? Have we the mindset, the wiring? Do we feel we have the resources? Can we design suitable equipment and facilities, not to mention institutions?

An answer we come up with could be that we are currently in crisis in all these respects. But might this be read as the Classical Greeks read *krisis*, a turning point as likely to be for good as for ill? As regards precautions being taken against our Earthly ambience being disrupted, these might help to bring the human race together, this with positive intent all round.

Extinctions

However, the ultimate possibilities are daunting. Thus nearly 250 million years ago, at what is thereby adjudged the end of the Permian geological period, vast volumes of low-viscosity lava spread across the surface of Asia eventually to cover 25 million square kilometres of land, with much ecological damage over and beyond . This event may have eliminated a good 80 per cent of the species then extant in the thriving kingdoms of animal life.

Just how big a part in life extinctions has been played by other such outpourings in locales around the world is a matter for debate. But at least seven out of 25 major waves of extinction in the geological record have been linked with evidence of large impacts by comets or asteroids.[2]

Volcanic eruptions can likewise be truly awesome phenomena with acute planetary consequences. Take the supervolcanoes that eject huge volumes of ash and lava from a caldera (a broad subterranean lava spread) as far as the global stratosphere. See Appendix B. The biggest the last two million years has been that on Toba in Sumatra *c.* 74,000 BP. Among other things, Toba brought our by then overly dispersed species dangerously close to extinction piecemeal. Much of the Yellowstone national park in Wyoming is underlain by a vast lava caldera that massively erupts every few 100,000 years or so. See Appendix B. In the highly connected world we are so carelessly creating today, such an eruption or, indeed, any large earthquake in a strategic location could be severely disruptive globally.

Tsunamis – earthquake or NEO shocks extending under water – often prove to be the most impacting of tectonic stresses – i.e. events caused or aggravated by natural faults and strains in the Earth's crust. They do this by generating seawaves which may be sustained a couple of metres or more in height across oceanic distances. The term "tsunami" is a Japanese one, its literal meaning being "harbour waves".

May one add that one feels somewhat uneasy about the dearth of general discussion of indications that the Earth's magnetic field is weakening and might flip into reverse soonish. What would be the physiological impact on ourselves and other animals?[3] And could our digital facilities be affected? Addressing this generic menace is hamstrung by uncertainties. Still, even if apprehension about magnetic reversal does come to seem only too well-founded, the said threat looks more liable to culminate in several centuries than within a matter of decades.

Information Flow

More immediately pressing and on-going is the need to keep in due perspective and adequately under control a communications revolution currently being led by the internet, the Worldwide Web and digital television. At Silicon Valley's Institute of the Future, Google has lately been characterized as "a religion posing as a company". And, since 2008 or so, there has been a marked upsurge of concern about how a novel imposition via the internet of consumerist values could be shaping, especially among the young, a lopsided and generally inadequate view of the world. In 2011, a Senior Fellow at the Roosevelt Institute warned that "in the end, a small group of American companies may unilaterally dictate how billions of people work, play, communicate and understand the world".[4] Might this eventually induce a reaction all round that is atavistically puritanical? Was it in some wise foreshadowed by the American elections of 2016?

One of the most concerned critiques to date came in 2008 from the then Director of the Royal Institution in London.[5] She saw the vacuity enveloping the young via the internet as part and parcel of a general loss of a sense of direction, a syndrome bound to culminate with the elderly: "Increasingly freed from fatal and debilitating diseases, they see decades of post retirement time stretching out to a solitary and empty end game" (p. 2). An especially detrimental side effect could be a diminished sense of where our individualities end and the outside world begins.

One is bound further to ask if there might not be totalitarian potential in much of what Susan Greenfield apprehended. In any case, how much commercial advertising do you wish your young children to be exposed to? How much sexual titillation or outright pornography? How much bedroom televiewing of whatever kind?

Education Structure

Polarization of the internet debate may still owe much to a continuing lack all round of hard evidence. But if education within primary, middle or high school is what one is most essentially talking about, the answer might be found in an inner core of time mainly given over to traditional teaching practice, this being enveloped by a well-netted outer zone. The latter would enrich the former. And for the latter, the former would proffer a solid foundation, a goodish puritan ethic.

Ambiguous Progress

It is, of course, by no means unusual for two contrasting perspectives to emerge in regard to technical innovation, be it broadly cast or very specific. The one may be that innovation will be unreservedly contributive to human progress; and the other that it may on balance be detrimental. As a rule, the verdict of History will be that each school of thought has had wisdom on its side. In some ways, change will have furthered progress; and in others compromised it.

Still, the rub throughout these days is that everything may be unfolding too fast to allow of well-considered outcomes in whatever direction. In 2012, the then executive director of Google opined that more new data was being generated every 48 hours than in the past 30,000 years. Moreover, quite a bit of this is destined to be mobilized, not just filtered out, in a world of rapid change and mounting complexity. Nothing is evolving more dynamically than information flow.

Paradoxical as it might seem, however, the overall impact, not least in the working environment, will be to generate insecurity along with a sense of sameness throughout. For out of the information revolution is dawning a great new wave of automation which will, within a very few decades, extensively displace human effort in many aspects of management, fabrication, transport and health. Too little attention has been paid as yet to the downside, most notably the qualitative underemployment.

Interfacing

Take surveillance in two of its many aspects. Remotely piloted helicopters, some barely the size of a clenched fist, may engage in reconnaissance, official or private. Electronic capsules will be inserted in arteries or veins for medical screening, diagnosis or therapy.

Then again, human decisions to act may indeed be expressed in electrical signals directly transmitted by wireless or by wire to a robotic arm. A decade back, Miguel Nicholeus of Duke University so stimulated the brain of a macaque monkey as to have it thought-control the movement of a robotic arm, mechanical transmission playing no part.[6]

Certain of the alternative applications will be military. As much was being noted in air defence half a century ago. A youthful brain takes 0.1 seconds to register a new visual image, plus 0.4 to interpret it. A further half second may then be required to activate the controls. The second thereby expended could be tactically excessive.[7] Halving it by exploiting the Hybrid Brain Machine Interface in the heat of battle should increase the attrition thus inflicted.

Ultimately, its application, aided by drugs designed to enhance cognition, could better be directed to helping in peacetime the disabled. Employed thus, it will represent a partial restoration of the natural order, not an aggravated departure from it.

In September 2011, a research report from the Berkeley campus of the University of California told of a surprisingly accurate registration, in motion and excellent technicolour, of what was then passing through an alert person's mind. And the previous May a Princeton report had similarly been tracking guinea pigs thinking about small objects from a collection randomly assigned to them. Here, too, the correlations were quite positive.[8] The rub has to be that such techniques might be exploited altogether too freely by an authoritarian political regime. Assured lie and uncertainty detection, in particular, may serve to undermine freedom more than to consolidate it. The same goes for speech recognition.

Furthermore, in any consideration of whether or how best to promote and preserve a world order comprising open and democratic polities, two paramount realities need to be borne steadily in mind. The one is that, on a planet shrinking by the day, every community (national, religious or whatever) is bound to become either a good deal more open and interactive or else a lot more closed. The other truism is that the number of talented and aspiring people coming regularly onto the jobs market will successively be liable to exceed, year on year, the number of challenging positions still waiting to be occupied, not least within the younger age range. A major contributory factor will be how the centralized control of big organizations is facilitated by the electronic revolution.

Further aspects can readily be cited of the automative outlook. Humanoid robots will help care for the sick and, in due course, advise and reassure. They may also collect rubbish, guard buildings and give street directions. Car driving on familiar routes will become progressively more automated. Train driving could become almost completely so, this side of 2050. Most of the male infants of my generation could not wait to grow up and drive trains.

Additive Manufacturing

Revolutionary fabrication possibilities are now being presented by three-dimensional printing, otherwise known as "additive" manufacturing. The 3-D machines, often of but desk-top size, will print out designs in metals, plastics, resins, woods and so on. Already it is possible to produce thus a variety of components which are (a) often accurate to within a millimetre and (b) readily adapted to customer requirements. Automated assembly has sometimes been feasible too. Modular building will likewise become extensive and customerized.

All in all, Information Technology is expected to affect the world of work every bit as profoundly as did the coming of the factory.[9] Granted, its innovations will sometimes call for decades of demanding development. At the same time, however, the distinctiveness of various more routine tasks will tend to diminish. Yet historically a singular working ambience has upheld many an individual's sense of identity.

According to Keynesian nostra, the structural unemployment engendered by IT should be resolvable by (a) shorter working hours all round and (b) putting substantially more resources into making ready, educationally and otherwise, for an age of leisure. However, there are always frictional resistances (institutional, cultural, individual . . .) to smooth adaptation. Moreover, the Keynesian assumption that there will be a simple 'multiplier' ratio between the level of new investment and that of other kinds of production in progress was never an eternal verity.

The crux will ultimately be less quantitative than qualitative. Loss of fulfilment may readily be incurred through the discarding of distinctive skills and singular responsibilities in favour of what may often be more commonplace IT. On present showing, those most threatened with actual redundancy near term will be lower-rank white collar people and the less skilled members of the blue collar community. Adequate remedies will not be forthcoming unless pertinent questions are frankly addressed in very good time, not least in government. There could be major implications for planetary stability and peace.

Refined Fascism

In a prescient statement penned in 1994, Edward Luttwak averred that the notion of market capitalism (unobstructed by regulation, cartels, trade unions, cultural inhibitions or whatever) as the "ultimate engine of economic growth" was now disputed only by "a few cryogenically-preserved Gosplan enthusiasts and a fair number of poorly paid Anglo-Saxon academics". Yet even when economic growth is low, this same capitalism will be liable to "inflict more disruption on working lives than individuals can absorb or the connective tissue of social bonding can withstand".

This prospect, Luttwak feared, could leave the field "wide open for a product improved fascist party" drawing much of its support (as per the 1930s) from a lower middle class losing out on social and economic security. However, he believed this stratum will need no longer to call up vulgar racism. Its declaratory aims, in so far as they can be foreseen, would principally be national economic sovereignty; limiting immigration; and winning, in immediate terms at least, the war against "terror".[10] But might not worse come later?

Scholarly Subcultures

Contemporary academe, whatever its overall pitch at a given time or place, readily separates into a broad assortment of subcultures. It is not just a question of Science versus the Arts or Classics versus Modern. Nor, indeed, of Physics versus Chemistry or Geography versus History. The specialisms, individual and institutional, can often be too widely cast so lacking in depth. And intercommunication can be insufficient to offset this. Climate change is a striking example of a subject that loses out on this account.

Faith and Life Science

A problem for the future role of religion/philosophy in human affairs is that life science impacts on it in two ways, these quite different but equally demanding. The one is indeed how relevant Consciousness may be to the concept of God within a cosmic or pan-cosmic setting. The other is the disturbed ecology of the tiny planet we find ourselves on.

Most of what we know thus far about our place in the total order of things has been garnered from this, our home planet. And not since Aristotle has the whole subject area advanced, on this basis, so fast as during the last half century or so. Witness the progress of human genetics. With 1954 came the determination of the double helical structure of the deoxyribonucleic acid (DNA) molecules which constitute the chromosomes bearing the genetic codes or genomes (i.e. assemblages of genes) that primarily shape on Earth all plant, animal and bacterial species plus certain viral ones. Complementary to, and historically prior to, DNA has been ribonucleic acid (RNA). A distinctive version of it, too, occurs in every living form; and its single helical structure leaves it very free to evolve. The enzyme from which RNA derives was isolated in 1960. In all the life forms we know of, DNA and/or RNA chromosomes are present in every cell.

Clinical Progress

The impact on human understanding of biological discovery gets stronger all the time. A signal aspect of late has been making more information available to patients, prospective or actual, about health prospects in general and for their individual selves in particular. Availability is mainly a function of the expansion of knowledge and resources. Yet it owes something to a burgeoning conviction that frankness not only facilitates professional-patient collaboration but may also diminish patient anxiety, this even in terminal cases. Medicine is becoming more and more an international culture.

Eugenics (1)

This line of enquiry should be able to reveal, as time goes by, the likely incidence of (a) some rare genetic disorder, (b) a more commonplace physical complaint like epilepsy, diabetes, breast or prostate cancer, a stroke or asthma, (c) mental instability, cognitive shortcomings, neuroticism, drug dependence or (d) such aspects of appearance as pigmentation, eyebrow type, bosom or height.[11]

How free should a medic be to decide how much to tell a couple? Should the partnership in question be thus assessed in terms of how firm their own relationship seems in other respects? Might a fascistic government – Blue, Red or hybrid – grasp such information in order to use it for social intimidation and control or, indeed, creation of a master race, however defined? And how much shall we want to know ourselves about our own futures? Want and not want? What will such knowledge do for social cohesion or, in the ultimate, for world peace?

Eugenics (2)

During the first half of the twentieth century, eugenics came quite widely to be seen as an operational option within polities ranging from fascist through social democratic to communist, its indicated aims ranging from racial purity to physical and mental enhancement. A common denominator very generally was a presumption that central government knew full well how its citizens might best evolve and therefore had a bounden duty to facilitate this. In Anglo-America alone, a good score of prominent personalities endorsing the principle of eugenics in some interventionist form or other included John Maynard Keynes, Linus Pauling, Theodore Roosevelt, Marie Stopes and H. G. Wells. The balance of this particular spread was well towards social democracy. Yet in the Anglophone debate, pre-1930 at least, the advocacy of eugenics became unpleasantly entrained with anti-Catholic and, albeit to a lesser extent, anti-Jewish hype.[12]

Demography

More immediately relevant, however, is recent experience as regards population growth. Here a salient truism is that the neo-Malthusian apprehension of but several decades ago seems to be turning into a self-defeating prophecy, on balance to date agreeably so. As recently as 1988, the then executive director of the United Nations Population Fund (UNPF) said she feared the 2100 world population total would top 14 billion.[13] Currently, however, received wisdom has it that the total will barely reach nine billion by 2050; and, a decade or so later, a gradual turn down will be getting under way. Especially critical in the West has been how Roman Catholic populations have been less and less inclined, the past century, to accept the Vatican's outright opposition to contraception.

Offsetting this now customary projection, however, are several more novel factors. One is a surprisingly strong tendency for life expectancies to rise, As everybody well understands, progress is continually being made right across medical science. Big advances in the treatment of various kinds of cancer are a salient example. One consequence is that, in countries with strong medical welfare throughout, the life expectancy of a healthy new-born baby will soon cap 100. What is more, within a few decades, genetic manipulation of stem cells – within a foetus or even later – may proffer the prospect of several more decades of life averagely.

This could also revive birth rates in that the virtual creation of an extra generation could induce young couples to widen the extended family base, in part to offset a decline in local community feeling. Meanwhile, individuals are getting bigger therefore hungrier, obese conditions apart.

Pandemics

On the other hand again, in August 2007 the World Health Organization warned that, thanks to globalization, pandemics could spread faster than ever before, biological warfare perhaps contributing. An on-going manifestation of the broad trend is how various diseases which modern medicine appeared to have routed for good have been fighting back. Tuberculosis and salmonella are well-known examples. So, too, is malaria, save that in this case the crux can often be the enhanced resilience of the mosquitoes. The zika virus gains likewise.

What is more, some maladies (e.g. SARS, ebola and HIV) would appear to have been afflicting our species for the first time, transferring in part from their customary hosts. Two consequences could follow. The one is that our immune system may not have had time enough to develop by trial and error an adequate response. The other is that so predatory a disease may be genetically less inhibited about inflicting high mortality if it does not ultimately depend on ourselves for accommodation. A bacterial collective unconscious may sometimes be at work.

Urban Living

A further caveat to enter is that commentators may have taken too little account of the part played in the pressure exerted on populations by rising prosperity and evolving life styles. Such oversight is liable to extend all too readily to such a consideration as the amount of extra living space the members of a rising generation may seek to have around themselves in their leisure hours in order to accommodate all their artefacts and aspirations, the latter including privacy and quietude. In Britain the last half century or so, the population has risen by a fifth but the number of households has by over two-thirds. The challenge thus presented, there and elsewhere, is all the more acute in urban areas with much less free space to play with. A basic standard of housing, within the context of precinct planning, should consistently be treated as a primary social need.

As late as 1800, just three per cent of the world's population lived in urban conglomerations. Today over half do. Moreover, the UN has anticipated that by 2020 the following conurbations will exceed 20 million: Delhi, Dhaka, Jakarta, Lagos, Mexico City, Mumbai, São Paulo and, far past that threshold already, Tokyo-Yokohama. An authoritative 2006 expectation had been that, between 2000 and 2030, the number of urban dwellers in the less developed or newly emergent world would double to four billion.[14] Our early hominid ancestors two hundred thousand years ago likely lived in tribal groupings a few hundred strong. How long can their descendants accept conurbations many millions strong?

China's Urban Crisis

With urbanization as with much else, China has evinced awesome ambition, not unaware that the outcome could be crucial to the entire world future. Her Prime Minister since 2013, Li Keqiang, is said much to savour a judgement by Joseph Stiglitz, Nobel Laureate in Economics. This was that technological innovation in the USA and urbanization in China would be "two keys" to global development in the twenty-first century. Mr Li himself has sanguinely described China's urbanization as a "gigantic engine" for growth.[15] Her defined urban population was *c.* 250 million in 1950 and 600 million in 2000. By 2050, it is currently expected to level out at a billion or thereabouts.[16]

Historically, large urban conglomerations have sometimes been especially prone to acute decline, given a situation in which a raft of negative trends (poverty; war; fire; undue migration, inwards or outwards; employment patterns; filth and disease; crime . . .) are mutually reinforcing. Already, China must have come fairly close to some such state of affairs in various of her burgeoning cities. In 2007, it was adjudged to have 16 of the world's 20 most polluted.[17]

The mammoth apartment blocks (said sometimes to be without lifts or escalators) can have about them a cheerlessness sufficient to make Britain's incongruent departures in that direction in the fifties look downright cosy. Much historic architecture has been demolished for redevelopment. Furthermore, a UN study has estimated that, across China, 400,000 are dying each year through air pollution principally. Far more than proportionally, these will be city dwellers.

Yet part and parcel of the current "Chinese dream" are some aspirational "eco-city" concepts. Witness the New Century Global Centre opened in 2013 in the south-western city of Chengdu. But one does well to remember that the "eco-city" school of thought emergent worldwide has been formidably lambasted for the perceived superficiality of its "neoliberalism".[18]

Far more dominant still in China's urban extension, in any case, are those mammoth sky-rise. And the outlook is for change coming slowly, as and when. But has Edward Gleaser of Harvard, a leading urban economist, been right to stress that, for budgetary and greenhouse reasons, city development should regularly be high density Chinese style?[19] Can that really be the answer for the old and the very young? Do "tall towers" cut environmental costs that much more than leafy suburbs? Would not residential low-rise with plenty of lawn be better adapted to changing needs and expectations, across a century and more? One suspects a lot of people would answer all, bar the last, in the negative. But perceived necessity short-term may impose "tall tower" policy choices.

Yet the recorded density of population within defined city limits is still relatively low in China: "Even greater Beijing, which city officials think is bursting at the seams, is far less densely populated than greater Tokyo".[20] However, this outcome has been sustainable thus far, there and elsewhere, by forcing 40 million Chinese farmers off nearby land over the past three decades, usually with little or no compensation. These seizures, which have ramified considerably of late, needs be legitimated and limited.

Meanwhile, a "nativist" hostility within the cities towards new rural migrants needs be curbed in both its societal and its bureaucratic dimensions. All of which will have to be achieved along with (a) an on-going reduction in corruption and (b) constraining the growth of an internal security presence said now to cost more than China's field forces and which is mainly concerned with inhibiting urban unrest, however this be expressed. As things stand, the indications that the Chinese urban crisis could acutely worsen look stronger than those which point to amelioration. There are simply too many conflicts of priority. And it may also be pertinent to recall that militant hostility towards cities as such has been generated historically in East Asia – the Huns; the medieval Mongol expansion; the Tai Ping rebellion in nineteenth-century China; Pol Pot in Cambodia of late . . .

Climate Change

A gentle warming may be maintained beneficially for centuries. But warming, regional or global, as rapid as what we have lately been experiencing can be very disruptive, not least as regards rainfall and snowfall patterns; and also be liable to drive up sea levels alarmingly, this well within a century. It therefore becomes a minus in planetary security.

Also to reckon with is the way polar ice sheets reflect the Sun's rays a lot more strongly than do ice-free landscapes or seascapes. One consequence is that, during times of global warming, temperatures rise more rapidly at high latitudes than near the equator. Shrinking ice and rising temperatures interact together. The temperature gradients from low to high latitudes therefore diminish.

All this effects some weakening of the "jet streams" that flow round the globe in middle latitudes a little below the stratosphere; and as this happens, their paths tend to become sinuously irregular. A consequence currently seems to be that much of the Earth's climate is becoming more erratic.

Western Europe has long been studied especially closely in connection with climate change this last millennium. The late Hubert Lamb, very much a pioneer of British historical climatology, read the northward limits of vine cultivation achieved in the said region during the High Middle Ages as betokening mean air temperatures in summer 0.7 to 1.4°C warmer than in the late twentieth century.[21]

The climb to that "little climatic optimum" (attained *c.* 1275 in west central Europe) no doubt owed a deal to the release of carbon dioxide through several centuries of aggressive deforestation in Europe and also East Asia.[22] The post-optimum temperature decline to 1750 (and the associated bad weather) can be understood largely, regionally and elsewhere, in terms of Milutin Milankovitch cyclicity – the rhythmic variations over thousands of years in our planets' orbit and axial inclinations.[23] But it has been attributed to vulcanism also. Moreover, diminished solar activity was also a significant influence later on.[24]

The underlying anomaly of greater winter cold within northern Scandinavia, as medieval warming built up generally, may have related to a polar anticyclonic circulation emergent as an offset to the incipient weakening of the winter Siberian high.[25] By 2015–16, there were tentative indications that much the same could be under way again.

Making past evidence of weather variations fuller and more definitive should further a trend within historical literature to take more account of impacting variations in the physical climate.[26] Recently, it has been proposed that mortality was so high in the Black Death spread of bubonic plague in the Europe of the 1340s (35 to 50 per cent of a population, as a rule) because debilitation was endemic, fairly hard upon a clutch of poor harvests with severe childhood malnutrition.[27] The plague had reached Western Europe after, in 1347, a Tartar army besieging the Genoese port of Kafka in the Crimea had catapulted infected corpses over the battlements.

The Tartar identity can be seen as a southerly and westerly derivative of the expansion under Genghiz Khan as from 1211. For several decades past, inferences have been drawn by those of us interested in how the little climatic optimum was made manifest in and beyond the Mongolian realm and the implications geopolitically.[28] Recently, a study in the *Proceedings of the National Academy of Sciences of the United States of America* has confirmed (via tree ring analysis) that abnormally warm and wet conditions in the Mongol desert heartland had followed upon years of drought. The grassland expansion that resulted allowed literally of the requisite horse power.[29] North-East Asia remains peculiarly susceptible to global climate trends.

Tropical Revolving Storms

Much later on in Mongol terms, Kublai Khan – a grandson of Genghiz – became the first emperor (1260–1294) of the Yuan dynasty in China. In 1274 then 1281, he dispatched big expeditions against Japan. The legend about the *kamikaze* ("wind of God") storms which wrecked each foray in turn is that they were conjured up by the Shinto gods revered at their shrine on the beautiful Ise peninsula.

The first of these meteorological events occurred that November; and there seems no good reason to see it as other than a strong winter gale, presumptively from the north-east. The second event, the decisive one, was something else: a terrific typhoon that reduced much of a Mongol-led fleet to matchwood. Moreover, it occurred on 15 August, well ahead of what today would be defined as Japan's typhoon season. The six most severe typhoons between 1945 and 1960 (as measured in terms of human fatalities) all struck her shores between 15 and 26 September.[30]

As the current debate about global warming has proceeded, something of a consensus has emerged to the effect that a sea surface temperature of 27°C is a threshold above which one is much more liable to see the generation of tropical revolving storms – the feature long known in East Asian waters as "typhoon" (from *tai fung*, the Chinese for "great wind") but elsewhere as "hurricanes" or "cyclones".

In the twentieth century, typhoons usually curving northwards towards Japan have originated very predominantly in the near oblong of Pacific ocean bounded by the parallels of 5 and 20°N and the meridians of 118 and 160°E: a sea area with a surface temperature often at or above 27° in late summer. An inference to draw is that seas even slightly warmer in high medieval times could have encouraged more and stronger typhoons, these occurring across wider seasons.[31] Cyclones have likewise been more active of late, not least around the more exposed shores of the Arabian peninsula. In November 2015, cyclone *Chapala* became the first hurricane-strength storm on record to make landfall in Yemen.[32]

Yet how readily this relationship will play out within the present climate crisis remains unclear. For one thing, the computers indicate that tropical revolving storms may not form that readily if wind speeds in the atmosphere are sheering too much vertically, a situation one presumes could be brought about or accentuated by rapid global warming. Even so, it is hard to believe that a global air temperature rise of two degrees or more would not find expression in a stronger tropical storm regime. Quite likely, this aspect of greenhouse will play in much more strongly over the next several decades.

Visibly affected climatically of late has been the semi-desert Sahel region of inner West Africa. A rise in sea temperatures in the Gulf of Guinea has apparently diverted rainfall southwards. Long term overgrazing and rapid population growth have compounded the adverse consequences. Witness the 2014 ebola scourge. Witness, too, how arms shipments from Libya to various Islamist insurgents have allowed of an active Jihad. Well over a thousand people have lately died in insurgent bombings in Nigeria alone. Meanwhile much livestock has been lost to hunger and thirst.

A Cult of the North

Across the whole field of climate studies, a disconcerting measure of uncertainty persists about the profiles of change but also about the human impact a given profile may have. And may not this impact in part be spiritual, not just economic or social in the regular sense? May not the disappearance of Arctic marine ice each summer (a prospect in train for several decades hence) have wider consequences sufficiently unsettling to induce or compound a spiritual crisis, one quite silent but very real?[33]

By the early nineteenth century the Russian tradition was being established which seeks to treat Siberia as an austere frontier within which the national spirit could eternally flourish.[34] In 1974, Alexander Solzhenitzyn, then in an unsettling exile, focussed on the "Russian North East" as "our hope and our reservoir".[35] But see Chapter 7(1).

Among other modern (say, post-1870) writers who have responded spiritually to high northern hemisphere latitudes are W. H. Auden, Jack London, William Morris, Fridtjof Nansen, Robert W. Service, Vilhjalmur Stefansson and (see Appendix B) Alfred Wegener. Do today's young people come to feel thus? Not so much, one senses, at present. But there could well be a strong global reaction in that direction if greenhouse warming slides out of control, as it could if President Trump finally rejects outright what he has called the "climate hoax". This could well prove to be the most critical test he faces.

Big Mammals

And is not concern widely aroused by the way we are still driving wildlife to extinction? Most conspicuously affected, yet again, are the big mammals. During the nineteenth century, the number of bison on America's Great Plains fell from five million to a few thousand. One recalls Chief Seattle's poignant address to the Dawanish League in 1854 as the United States expanded West. He told of the "great sadness" that would fall upon the Earth once all big beings had gone.[36] Keeping a few survivors sadly inactive in zoos will do little to mitigate this. We may also be threatening insect life far more than we recognize. Are these not matters for planetary action more forceful and co-ordinated than anything we have yet seen?

Nuclear Need

Warming consequent upon the man-induced build up in the atmosphere of "greenhouse gases" is quite the most enveloping threat we currently face environmentally. What needs be guarded against, in this respect, is adopting strategies of threat mitigation that could create fresh problems comparable in scale.

Take the menace that would be posed by the high atmospheric pall created by an asteroidal or meteoritic strike or maybe a supervolcanic eruption or, indeed, a heavy military nuclear exchange. In the aftermath, the "renewable energy" options much in current favour as "green alternatives" (solar cells and wind farms) would lose much of their operational efficiency.

This is because the Sun's input directly to the solar cells and indirectly to the wind farms could be acutely compromised by the same pall as was driving ambient temperatures down across our planet. The zonal distribution of wind energy would thereby become subject to continual chaotic change. Moreover, the large-scale storage in advance of electrical energy still looks problematic, recent progress in this direction notwithstanding.[37] All of which reinforces, to an extent, the received case for some retention of nuclear power, hopefully under regional or multinational management. Other potential bottlenecks in global resource supply within several decades are in sea-fisheries and certain non-ferrous metals. Of the last, lithium for rechargeable electrical batteries is a salient example.

Social Crisis

All too many criticalities in resource management can be identified for the decades ahead. In his seminal 1963 essay on the ambivalences of economic growth, the late Mancur Olson stressed a general failure to appreciate how often this created more losers than winners.[38] Since when, the digital revolution in Information Technology has been exposing decision-taking in many fields progressively more to public view. At the same time, however, it has been concentrating actual policy decisions in fewer hands.

However, what needs to be focussed on more specifically at this juncture is how the narrowing apex of really challenging jobs stands in ever starker contrast with the broadening profile of people (especially young people) keen and able to do them. Very many persons yearn to make a distinctive contribution to the common weal over and beyond their interaction with family and close friends. If they are effectively excluded at national or regional level, they may seek to do more locally or simply through the internet. But this way through will need careful scene setting to be really contributive. It will have to involve a major reconsideration of applied philosophy, not just of rickety pragmatism nor of econometric profit, too narrowly defined. Taken as a whole, young people are not looking, first and foremost, for "happiness" defined in consumerist or, indeed, spiritual terms. They are looking for constructive fulfilment. Not finding it may all too readily lead to alienation, passive or militant.

Conflict Management

So what of the starker alternative of angrily venting self-expression in ways negative and maybe extremist? One thinks, *inter alia*, of the more *ersatz* and fantastical of the many new religions or cults now flanking traditional beliefs; the promotion of, say, astrology or homeopathy; and, most particularly, "end of the world" religious visions or conspiracy theories. Such tendencies may further the splintering of mass society making it all the more vulnerable, from without or within. Meanwhile there may now more often be something veritably ideological about the sheer selfishness of those who turn gratuitously to crime, be it gang or lone wolf.[39]

An overview from the Institute of Psychiatry at King's College London has identified several factors as contributing signally to what the authors see as a rising incidence of paranoia worldwide. Those they select are migration from country to country; migration from country to town; urban loneliness; and a pervasive sense of victimhood.[40] The question that will always arise is how readily may the said syndrome be productive of alienation, either passive or radical.

These last few years, the information revolution may have enabled governments to curb dissent more effectively, this more or less regardless of whether they have bothered much or at all with the broader policy accommodations previously seen as essential to securing law and order.

However, this state of affairs may be starting to change. Not least is this because the threat of biological warfare looms on the horizon. It is a threat which subversives could be all too well placed to exploit, especially those so imbued with political or religious millenarianism (or simply purblind anger) as to be prepared to strike suicidally.

Much of what has just been said could apply as well to confrontations between two standard armies. In such scenarios, however, each side may have become very uncertain about the specific preparations needed to ensure contemporary deterrence or to wage modern war: what mix of weaponry; what force densities; what surveillance; what command and control; and what logistic support. So what about arms control? And what of cyberwar as a novel dimension? Chapter 8 aspires to help draw such questions into the wider armed conflict debate.

All in all, we could be said to be moving into a world situation too fluid and complex to manage via customary modalities. But might the evolution of a new philosophical perspective, planetary and cosmic, help to bring things more into focus? And might our scientific and philosophic backgrounds be harmonized to this end? Above all, though, have we the time left for such a consummation? Could a remit to cope with mega-natural disasters help to consolidate a global regime? And could it proffer planetary governance of the singular character required?

4

Planetary Confinement

(1) Limited World War?

High in my appreciation of the visiting fellowships it has been my privilege to experience has been that in 1980 to the Australian National University in Canberra. It was held for a single term at the Strategic and Defence Studies Centre within the Research School of Pacific Studies. At the time, Australia as a whole was becoming more conscious of its pivotal role as an Asia-Pacific power. Directing the strategy programme was Robert O'Neill, a Vietnam War army veteran who went on to be Director of the IISS and then Chichele Professor at Oxford. His has been very much a voice for moderation in world affairs.

A Canberra Perspective

My assignment became the writing of a monograph we decided to entitle *Limited World War?* It would explore the notion that a nation or movement possessed of extended geographical reach will be the more liable to take dissuasive or punitive measures against specific adversary action; and may well do so in a different part of the theatre or of the world.

It was seen as a pattern of conflict most likely to occur, to judge from the then recent past, between major powers or alliances locked in near parity as regards stable deterrence at the top strategic level. Lately, the plots have got more complicated. In the Ukraine, Russian minority separatism has been encouraged forcefully from Moscow. In Syria, the Russians struck more at moderate than at extreme Islamic insurgency whereas the West sought to intervene militarily just against the fascistic "Islamic State" hard core.

Meanwhile, Islamic insurgent radicalism (El Qaeda or Islamic State or eclectic) has turned actively aggressive across much of Eurasia and

Africa, not to mention the USA. Both Western and Russian commentators are referring to the present pattern of insurgent-cum-regular local conflicts with global import as "hybrid" war.[1]

The 1967 Crisis

There are lessons to be learned from closely comparable episodes during the Cold War era. Thus it is conceivable that a wish to divert the USA from a Vietnam war that still hung in the balance lay behind the curious machinations by Moscow in May 1967 apropos the Arab–Israel conflict. Yet this is by no means to say the Soviets sought the all-out war which resulted from their propagating the fiction that Israel was making ready for a major strike against Syria.

One evening that January, as a journalist filing for the *New Statesman*, I had had an extended discussion in Cairo with a senior affiliate of the Nasser regime. He was categoric about three points: (a) another war with Israel was inevitable, (b) when it came, the Arabs would win, whatever America or Britain did, but (c) this war would not be launched for several more years. This deferment was, he said, because Egypt needed time to wind down its heavy involvement in the Yemen's internal strife but also to get a large-scale birth control programme under way at home. He said nothing about the terms of a post-war peace settlement. One should have pressed him on this.

Five months after that interview, the outbreak of the June war showed how the USSR had underrated the credulity of Egypt's President Nasser and the impetuosity which stemmed from it. Nor may she have allowed sufficiently for the concern felt by King Hussein of Jordan that, if he did not join the Arab military compact hastily being put together by Cairo and Damascus, he would be overthrown by Palestinian insurgent radicals within his kingdom.

In fact, the converse happened when the royalist versus insurgent showdown finally came in 1970.[2] The Palestinian guerrillas fought hard but were defeated in just over a week. A primary reason was starkly obvious. The insurgents had not been given by their main arms suppliers, the Soviets, a proper anti-tank capability. This would have rendered them too autonomous for Moscow's liking.

Europe as Hostage

Something the USSR was always deemed to be appraised of post-1945 were the hostage connotations of Western Europe in general and West Berlin in particular. A primary manifestation thereof was the wholely disproportionate fraction of the Soviet thermonuclear missile force that was emplaced inside the USSR's western borders, targeted on NATO Europe. More overt, though probably tokenistic, had been Moscow's top-level warning during the Suez crisis of 1956 about rockets being fired in retaliation at London and Paris. Further, Nikita Khrushchev alluded to a possible Soviet invasion of Turkey, then very much a Cold War ally of the West.

As regards West Berlin, one was always told in NATO that Soviet interference with that truncated city could precipitate general hostilities. So war because the Warsaw Pact was blocking the access autobahns awhile? This was none too credible as expressive of deterrence. Nor was the West's successful use of airlift to break the Soviets' blockade of West Berlin (1948–9) all that reassuring a precedent. The respective implications as time went by of two variables were and remain uncertain. The United States was then bound to lose soon its military nuclear monopoly. Meanwhile, runway dilapidation in West Berlin was becoming critical.

May we consider, too, a relatively benign turn in Cold War diplomacy. In October 1979, the USSR announced a unilateral withdrawal of 20,000 troops and 1000 tanks from Ground Forces Soviet Germany. In other words, its standing army there was being reduced from 20 to less than 19 divisions, the first time its order of battle had been altered either way in a quarter of a century.

With this adjustment, Moscow was signalling the interpretation it wished all to receive of the full-scale invasion of Afghanistan it was covertly preparing to launch that winter. This signal was that the move should not be seen as the prelude to a more general geopolitical expansion. Rather, it was merely intended to prevent the fall of an undeniably unpopular Marxian regime in Kabul. For this might generate a "domino effect" within the constituent Moslem republics of the USSR itself. A decade or so later, this much was to happen regardless.

(2) Ambiguous Togetherness?

Limited World War was a simple model addressing what is liable to be an intricate pattern in world developments overall. An alternative way of conceptualising pressures we might readily be subject to is to see them as essentially a function of a "hit the buffers" effect: the way human progress can quite suddenly seem to be bounded more acutely by the limited dimensions and potential of our home planet. Various material constraints and/or cultural developments can induce this impression in contemporaries, sometimes with negative implications for harmony and peace.

Although we were not subject to too much in the way of natural disasters in the Eurocentric world leading up to 1914, apprehension of those approaching buffers was increasing, not least on account of how readily one warlike crisis was succeeding another as calamity closed in. Yet strong, too, was an enduring belief that the post-1815 Concert of Europe tradition (dynastic marriages, naval reviews, commercial exhibitions, professional diplomacy, *ad hoc* multinational accords . . .) could contain tensions adequately, a while longer at least.

Subsequently it was inferred that "the little Eurasian peninsula that was Europe, which had conquered the world and was its powerhouse, contained too much energy and power for the narrowness of its confines. The very process of imperial activity had simultaneously furnished occasion for clashes and crises; and served in the function of safety-valve for the overflowing energy of Europe. There was, in 1914, no more room left for fresh conquests".[3] Nor was there much for other manifestations of aggressive nationalism, this sometimes promoted as a welcome alternative to philosophies more Leftist.

Awareness had become acute that, simply in geographical terms, bounds were being reached on all sides. So many events spoke for themselves, In 1909, Robert Peary became the first explorer to reach the North Pole.[4] In 1911, Roald Amundsen was first to the South Pole. In 1906, Amundsen had conquered the North-West Passage from Arctic Canada to the Pacific.

In 1890, the "moving frontier" of the Old West had ceased to figure, as a special though ever shifting territorial zone, in the US federal census. Likewise, in 1886, the first Canadian Pacific train had run from Montreal to Vancouver, thereby fulfilling a precondition of British Columbia's accession to the Canadian confederation. Russian promotion of the Trans-Siberian railway had been drawing on this precedent repeatedly since 1879.[5] And by 1875, the exploration of Australia's arid western plateau had become but a dusty "mopping up" of isolated tracts. The formation of the Commonwealth of Australia in 1901 set the seal on the continent's being well and truly spanned.

The European political impact on Africa evolved from being peripheral to comprehensive between 1881 and 1911. Similarly, in inner Asia, a British military expedition reached Lhasa in 1904, the upshot being that in 1912 Tibet became autonomous within China though under British protection. Worthy of mention, too, are Sir Aurel Stein's expeditions deep into inner Central Asia between 1900 and mid-1914:

ventures primarily undertaken to explore the roots of oriental civilization but with contemporary geopolitics also in mind.

Communication Diversified

The flow of people, goods and data was transformational the while. The switch from sail to steam at sea was virtually concluded. A global network of underwater cable links was effectively completed with the laying, from 1902, of a bifurcating line from British Columbia to Queensland and to New Zealand. And in 1901 Guglielmo Marconi (1874–1937) had made his pioneering wireless transmission from Cornwall to Newfoundland.

The oil-driven internal combustion engine was now figuring on the surface and in the air. The first petrol-driven cars had appeared in Germany in 1885–6; and Henry Ford's Model T made its debut in 1904. Tentative experiments with heavier-than-air flying machines had begun in the 1890s, Orville (1871–1948) and Wilbur (1867–1912) Wright made, in 1903, a very short but well controlled flight in a plane they had made themselves. However, internal combustion was to become, in that first generation, not so much the supreme herald of freedom and peace as the ultimate fount of war.

Being close to the present day, this era extending through the turn of the twentieth century proffers the more pertinent cautions about the ambiguities of progress, especially in communications, literary and transport-wise. However, its lessons are reinforced by two preceding eras in the history of Europe in particular. One has in mind the Renaissance and the Enlightenment.

The Renaissance

The delimited yet ongoing humanism of the Renaissance (in the fifteenth and sixteenth centuries) was eventually to be compromised considerably by the brutality of religious strife. Europe's movable-type adaptation of China's woodblock printing was sadly conducive to this outcome, especially in Germany.

During the second half of the last century, a succession of social philosophers and historians tasked themselves with interpreting, sometimes in terms of psycho-history, the great Renaissance paradox. An overview of their insights is essayed in *The Bounds of Liberalism.*[6] Suffice for now to note the gist of Bertrand Russell's perspective. In the Middle Ages, "great originality was scarcely possible. The Renaissance restored human pride but carried it to the point where it led to anarchy and disaster".[7] Copernican astronomy failed to prevent this culmination; and arguably, indeed, accentuated it. Much the same might be said of the maritime voyages of discovery.

The Enlightenment

Between the Renaissance and the present day stands the late seventeenth and eighteenth century Enlightenment, a less remarked but very seminal ascent of creativity within the culture of the West. Its precursor was a great revolution in mathematical physics, this spearheaded by how Isaac Newton recast and codified extensive European thinking about gravitation. Newton himself confessed to being mystified as to how a body anywhere in the universe could exert a pull on every other one wherever. But he insisted that the prior question had to be whether refined observation and calculation confirmed a correlation. They did, and have ever since, subject only to Einsteinian qualifications feeding in at extreme values.[8]

Though American, Scottish and English thinkers were prominent within it, the Enlightenment was led largely by French thought. It can duly be seen as a precursor to the French Revolution and the Reign of Terror. Whence it led on continentally to the aggressive ratiocination of Napoleon.

This sequence could have culminated much more negatively than it actually did. For Napoleon's remake of a formidable French army in the months after his escape from imprisonment on Elba cannot but be seen as one of the greatest triumphs for personal charisma in the whole of recorded history.[9]

Accordingly, it could readily have led to the crushing of Wellington at Waterloo. The Bonapartist army might well have prevailed on the day had Blucher and his Prussians not arrived just in time to lend Wellington critical support against the tiring French. Yet had Napoleon triumphed at Waterloo, his ravaged army would probably have been overrun by the other allied armies convergent on it once again. At all events, the most likely outcome overall would have been a Holy Alliance comprised of Austria, Prussia, Russia and likely France upholding reactionary absolutism across continental Europe for decades to come.

Our Present Discontents

Transposing this line of argument to our present discontents, one observes a great intellectual revolution getting under way – notably in information science, philosophy, biology and astral physics. Again, we cannot be sure where or how far it may take us. We therefore do well to develop any routine which, in addition to its intrinsic benefits, may enhance our sense of planetary togetherness.

Defence against NEOs and other natural hazards could hold the key to this. It is submitted throughout this study that the military thermonuclear powers are the best placed to take a lead in this regard. But an analysis launched in 2006 well stressed that, for many other purposes, Brazil and India at least can be seen as emergent great powers, particularly in view of "dramatic changes in the character of international society – the exponential increase in the number of international institutions and in the scope, range and intrusiveness of international rules and norms".[10] What is seriously lacking at present, though, is an Islamic nation ready to assume great power status. Neither Turkey, Pakistan, nor Indonesia appear to be. Nor Bangladesh thus far. Yet a Limited World War is currently in progress between radicalized Islam and many of the polities rooted in customary Islam or Christianity. This is clearly inimical to a planetary perspective.

A question arises as to whether India, Hindu yet polycultural, could have a special part to play in closing divides. May she be able to do so as her economy gets stronger? Will she be able to if, as seems likely, religious ties progressively weaken further and turn less exclusive? Will she be able to if the world as a whole goes ecumenical in a very full sense, not just as between Christian denominations but between all theologies and philosophies and, indeed, natural science – cosmology prominently included?

Provenance could here be provided by Mahatma Gandhi (1869–1948). He may not mean a lot to the policy makers of modern India. But he was very much *primus inter pares* (in India, Britain and beyond) for the generation or two ahead of mine. He became a singular subject of contention, much of it acerbic though some too indulgent. But there was a kind of moral strength in his disinclination to decide whether his outreach was essentially to lower caste Hindus or to humankind in general.[11]

What is needed in the current fluidity is enough empathy with Islamic history and aspirations to bring the Moslem mainstream more fully on board. This must involve self-determination for the nationalities involved and fulfilment for their scores of millions of individuals. Otherwise, the "so-called Islamic state" propensity will assume still uglier forms, Islamic or whatever.

The threat posed basically emanates from the Assads of this world though also from the Trumps. Action and Reaction is still working out in a manner too negative for future planetary security. Yet modern India may have a valuable bridging role to play. Originally, after all, the subcontinent did much to spread the foundations of culture eastwards and westwards in respect of such basics as written languages[12], arithmetic notation, yoga, chess and astronomy. And the religious plurality that may weaken modern India geopolitically may come to strengthen her philosophically. The fascination with astrology she shares with other Asian countries may give way to a more cogent concern about our planetary situation. And out of this may come a singular contribution to the anti-dogmatic inter-faith planetary ecumenism that *inter alia* the Dalai Lama and Britain's lately retired Chief Rabbi have called for. But nothing is guaranteed. Ever since the 1947 partition of the subcontinent, the governance of the republic of India has been subject to a strong anti-Islamic bias.[13] And Pakistan could well turn unstable.

The science-based information explosion now gathering strength globally means that the world of the next generation will have to be throughout either a sight more liberal or else a lot less so. The former outcome will not be easy to ensure.

5

The Abrahamic Schism

(1) Islam

Western commentators sometimes remark on how Islam originated six centuries after Christianity, the connotation being that its collective persona remains more youthful, therefore more forceful.[1] This may be unsound. An organic model with built-in growth and ageing hardly befits a religious or political movement.

A more instructive point of departure could be that, in the Arabian birthplace of Islam, humankind has stood more exposed to water shortages than has been the case almost anywhere else, this largely due to marginal and irregular climate dynamics. Thus in 540 AD the Krakatoa volcano (close by Sumatra and Java) erupted with more force than has any other the last 74,000 years: that is to say, post-Toba. In a 1999 pioneering study, David Keys provisionally related this causally to disturbance near-term in the water cycle in Arabia Felix: the relatively salubrious south-west of that wide peninsula.[2] This well-defined region had long been a node in Indian Ocean trade, notably as a leading producer of frankincense and myrrh. A kingpin of its economy was a stone faced dam (600 metres wide) close by Marib, the capital of Sheba.

During a flood in the 540s, this dam was breached for the first time since 450 AD. A decade later it happened again. Following a third breach in 590, the weakened dam was abandoned. An allusion to its demise can be found in the Qur'an, the inference therein being that the locals had been turning away from Allah.[3]

Emigration from Arabia Felix duly got under way. Not least was this to Yathrib, an oasis and the second city in the central Hejaz. Already the whole region was in the throes of a water shortage extending long term.[4] Obversely, since the third century the Hejaz, led by the Quraysh tribe, had experienced a cultural uplift. By 600, an Arabic script will have been set down. Witness, too, a collateral Quraysh trend away from single-tribe gods and towards a broad monotheism. The black meteorite in the Kaaba at Mecca was readily seen as connoting a universal deity.

Into this milieu Mohammed was born *c.* 570 AD. Orphaned in infancy, he grew to well-rounded manhood. At 25, he embarked on a happy monogamous marriage with Khadijah, a wealthy Damascan widow. Though said then to be 40, she went on to bear him six children.

By 610, he was experiencing, through cave meditation, revelatory visions. As to whether these connoted a deep mental disorder, this comment by an ordained Christian priest (and one time Professor of Arabic at London then Princeton) is pertinent. Those who question his "stability do so only by ignoring the overwhelming evidence of his shrewd appraisal of others and of the significance of what was going on in the world of his times; and his persistence in the face of constant opposition".[5]

Three years after his first revelation, Mohammed emerges convinced he is the messenger of Allah, the one true God. Though his witness was to cover life in every aspect, emphasis was placed throughout on *zakat*, charitable expenditure on a regular basis: "For your kin; for orphans; for the needy; for the wayfarer; for those who ask; and for the ransom of slaves" (*sūra* II v. 177). It is appropriate thus to cite the Qur'an, since it was declared to be a compilation of the Prophet's oral pronouncements. A full text was generated the next two centuries. To devout Moslems, it has always been an unerring numinous authority.

A millenarian climax is therein foretold. It will usher in a final Day of Judgement which will utterly transcend the Lesser Judgement effected at an individual's death as well as the awakened Inner Light a soul may experience at any moment.

At this fell time, "the sky is cleft asunder . . . the stars are scattered . . . the oceans are suffered to burst forth . . . " (*sūra* LXXVII, v. 1–4). All souls are then weighed on the scales of justice for their every thought and action on Earth; and are duly consigned to either Heaven or Hell. The former is foreseen as untrammelled relaxation within a lush Garden of Eternity. The latter is portrayed as ferociously fiery – furious flames, bubbling pitch, boiling water. . . . "Those thence consigned will dwell therein forever. Except as God willeth" (*sūra* VI, v. 158). This stark divine message was relayed by a Prophet from then on sublimely convinced of the authenticity of its transmission via the Archangel Gabriel.

To say his vision went down ill with the Mecca establishment would be to understate. His insistence on the voluntary redistribution of income hit many pockets. So, too, did his condemnation of the idols which pilgrims had paid onerously to engage with. Undaunted, he consolidated a cadre a few dozen strong. Beyond that, progress was, to start with, hard to register.

An opportunity for break-out came from a Yathrib then torn by inter-tribal and anti-Jewish violence. A group of locals, impressed by Mohammed's personality and prophecy, invited him to help them build social peace. Yathrib, now renamed Medina, thereby became sufficiently secure a base for warlike operations against Mecca, these starting with caravan raiding.

Come 630, he was able to return home in almost peaceable triumph. But it was to a Mecca that, the past year or two, had been plagued by aridity so bad that (according to Arab sources) the people were turning to cannibalism. No doubt the warmth of his welcome (and then the explosive spread of Islam to embrace half of Arabia in what proved to be the remaining two years of the Prophet's life) owed something to a concurrent easement of the droughtiness.[6] Then as now, rainfall fluctuations across the Arabian peninsula will have been very localised.[7]

A Bedouin Faith?

It is tempting for westerners casually to see Islam as a bedouin philosophy evolved under desert starfields and overlooked by resplendent desert Moons. However, neither the Holy Qur'an nor modern scholarship encourage this characterization. Not unrepresentative a negative citation is one which says the "Arabs of the desert are the worst in unbelief and hypocrisy, and most fitted to be in ignorance" (*sūra* IX, v. 97). Likewise, one Islam scholar has seen the faith as able to "multiply cities at will".[8] Another stresses how the location of the central Hejaz (plus the way the winter rains had usually ensured groundwater enough there the year round) had created possibilities for trade, hence a measure of urban growth.[9] Nor should one overstate how crucial bedouin martial prowess was to spreading the faith. Mounted warriors a few thousand strong could not have seized and held big cities had strong local influences not worked in their favour.

Something that plainly emerges from Mohammed's witness as recorded in the Qur'an is how unacceptable idolatry is in Islam, a reaction evidently related to an emergent taboo against imaging Allah. As part and parcel, local *sūfi* saints of Islam were initially viewed with unease, interposed as they were between the laity and Allah. Nevertheless, a proliferation of sainthoods occurred as Islam extended and consolidated.

The Qur'an

Ninian Smart has put to us that to think of it simply as "a holy book . . . does not convey the centrality and power of the Qur'an in Muslim eyes. If you were to look for a rough equivalent to the Christian Incarnation, . . . it would be the Qur'an . . . as divine speech that lies behind and within the Qur'an is eternal. All this gives it a much deeper meaning and a much higher status than that of the Bible".[10] Mohammed was, after all, the final Prophet – he who had presented the complete and definitive account of Allah's revelation.

A most basic question to ask is how progressive the Qur'an's message was in relation to its day and age. In those terms particularly, it comes across as enlightened in certain cardinal respects. The injunction "Kill not your children for fear of want" (*sūra* XVII, v. 31) took courage to enunciate at a time of acute aridity leading, it seems, to infanticide and maybe cannibalism. As much goes for the liberal rules proposed for slave manumission; and the proscription on forcing female slaves into prostitution (*sūra* XXIV, v. 34). Likewise, the status accorded women (*sūra* IV, sections 1–6) may still seem retrogressive from modern perspectives. Yet to quite an extent, their pre-existing status was thereby consolidated or enhanced, institutionally as well as in regard to codes of male conduct.

To my mind, a most edifying aspect of the Qur'an concerns the references made to animals. Though few and brief, they explicitly accord every member of the animal kingdom an independent place in the scheme of things: "the birds of the air with wings outspread. Each one knows its own mode of prayer and praise. And God knows well all that they do" (*sūra* XXIV, v. 41). Indeed, all animals live in communities just as we do and "shall be gathered to their Lord in the end" (*sūra* VI, v. 38).

These are beautiful sentiments, beautifully expressed. Would they have gained traction in the non-Islamic world, rather than so much being made of Qur'anic allusions to war-waging. Not that these actually are so very extreme. There is recognition of a contingent need to "Fight in the cause of God"; and on behalf of those who, being weak, are ill-treated. Granted, these criteria are not always well spelt out. But a general understanding would appear to be that (a) war should normally be just for self-defence and a restoration of the *status quo ante*, (b) women, children, the elderly and infirm should not be molested, (c) trees, crops, and water sources must not be destroyed and (d) peace should not be withheld once an enemy is willing to come to terms.[11]

Common Heritage

Often remarked are those materials in the Qur'an which tell of a common heritage with the other two great monotheisms that developed on the western fringes of the awesome Arabian desert. In the Qur'an, Abraham is cast as a very archetypal desert father, committed to the one true God. This allows that one should interpret the designation of the Ka'aba at Mecca as the "station of Abraham", understood to have been created by him and his son, Ismael (*sūra* II, v. 125–7).

Meanwhile, Adam is presented as a truly universal spirit, one edified to boot. In the Bible, he has been given charge of the Garden of Eden. But through the Qur'an, Allah makes him "vice-regent on Earth" (*sūra* II, v. 30), someone the angels must obey. This elevated designation lessens the significance of expulsion from Eden, an imposition in any case described therein as temporary (*sūra* VII, v. 24 and *sūra* XX, v. 129). Absent is the Biblical disposition to cast the eviction of Adam and Eve starkly, a pointer to Original Sin. Instead, Adam is God's "most important creation".[12]

A critical test of the prospects for concordance with Christianity must be the Qur'anic coverage of Jesus and his circumstances. Mary is told by God of His ability to will a virgin birth (*sūra* III, v. 47). Jesus Himself is accepted as "a righteous prophet" (*sūra* VI, v. 85) whom God raised up "unto Himself" (*sūra* IV, v.158). But there a line is drawn. He was never crucified by the Jews (*sūra* IV, v. 157); and so presumably experienced no bodily Resurrection.

Nor should He be considered part of a three-in-one headship. Addressing Christians as "People of the Book", the Qur'an advises them to "Say not 'Trinity'. Desist. It will be better for you. For God is one God. Glory be to him" (*sūra* IV, v. 171). After all, God could not have had partners in Creation; and so would have had none since (*sūra* X, section 4). Yet despite their aberrant understanding in this respect, God has instilled "compassion and mercy" in the hearts of those who follow Jesus (*sūra* LVII, v. 27).

The Nature of Allah

Still, the main import of the Holy Qur'an concerns the nature of Allah. A cardinal axiom is that "No vision can grasp Him but His grasp is over all vision" (*sūra* VI, v. 103). This locates Him much higher above us than the loving Christian Father or than Yahweh, God of Israel. Though He may seem nearer to Man "than his jugular vein" (*sūra* 1, v. 16), this simply follows from His insight into the whole of Creation.

Humankind may be comparatively close to Him though only in that He has set us above all earthly beings. Allah does hear our prayers but can only respond proportionately. To ask for more would be to "transgress beyond all bounds through the Earth" (*sūra* XLII, v. 27).

As Allah metes out justice, He is eminently fair, totally informed and prepared to reconsider should a culprit repent. Yet in the final analysis, Allah is contingently prepared to destroy a whole people "when its members practice iniquity" (*sūra* XXVII, v. 39). This harsh austerity is further much manifest by His detachment from familial concerns. He has no consort. He has progeny only in the ultimate sense that every living being is within His scheme of things (*sūra* VI, v. 107). He appears less hard to reconcile with cosmological perspectives than do most of the supreme deities humankind has identified.

To its believers, Islam basically means submission to the will of Allah. What Moslems reject, however, are "most malicious" suggestions that this connotes fatalism and resignation. We are advised that were "man not to be free, religion would have no meaning".[13]

Out of Arabia

To those at the receiving end, the Moslem expansion out of Arabia will have come across as a cadre one, extending a new religion and language, not an entire people. On that basis, Damascus and Jerusalem fell in 635 and 637 respectively while Alexandria was to in 642. The advance along the North African coast to Tripoli took but two more years. From Gibraltar to Toulouse under ten.

An eventual Islamic conquest of Western Europe was finally precluded by the Frankish victory in the Battle of Tours on the River Loire in 732, just a century after the Prophet's death. In 759, the Moslems withdrew for good behind the Pyrenees.

On the eastern flank, Basra fell in 656, Kabul in 664, and Bukhara in 710. The involvement of Islam in India likewise began in the seventh century AD. Then under its fifth Abbasid caliph, Harun ar-Rashid (ruled 786–809), Baghdad became "the richest city in the world. . . . Arab merchants did business in China, Indonesia, India and East Africa. Their ships were by far the largest and best appointed in Chinese waters or in the Indian Ocean. Under their highly developed banking system, an Arab businessman could cash a cheque in Canton on his bank account in Baghdad. In Baghdad everything was plastered with gold".[14] By 1600, Islam was the dominant religion within the Indonesian archipelago. By 1100, the West African kingdom of Ghana had been taken over by an Islamic confederation.

Widely across the Fertile Crescent from Palestine round to upper Mesopotamia, there is evidence (e.g. in church locations, olive planting and agricultural terracing) of an upward rainfall trend peaking out in the sixth century. For the Nile valley, the evidence points to a similar conclusion. Likewise, Cyrenaica was manifestly drier than in Graeco-Roman times. What has not been demonstrated, it seems, is that the initial expansion decade, 632–642, was especially characterized by climate adversity throughout the region impacted on.[15]

Divine Unity

Mohammed may have found little philosophic depth among the bedouin. Nevertheless, Arabists in the modern West have tended to credit those they observe with much spirituality. An enduring "noble savage" strand within the literary desert romantics of Britain and France could be one explanation. But among the others can be that a spiritual expression does not necessarily involve elaborate philosophical preparedness.

In his 1997 study, Seyyed Hossein Nasr spoke of desert dwellers "great love for Nature . . . a deep intuition of the presence of the Invisible in the visible".[16] On the face of it, this suggests a godliness more immanent than some interpretations of Allah. Absorbed into Islam early on, this will have paved the way for the precept of indivisible understanding. All thought is thereby open to religious infusion.

From whence arose the *madrasah*, a "place of learning" usually located within a mosque complex. Rather over a millennium ago, this concept was developed further with centres of higher learning teaching subject mixes ranging from religious law to astronomy. A hospital might well be attached.[17] This pattern was to serve as something of an initial model, from the twelfth century, for university development within Catholic Europe.

By then, however, Islamic academe was turning defensive, fearful lest doctrinal ructions furthered a disintegration of *dār al-Islām*, the Islamic homeland as lately established.[18] Intrusions by Crusaders and then Mongols compounded the apprehensions. Moreover, it proved impossible to accommodate evenly an accelerating growth in knowledge diversity. The upshot was inhibition all round. The *madrasahs* duly turned inwards.

Still, Islam's territorial extension had encompassed major centres of scientific enquiry, enabling these to associate more readily. Not least could this apply to schools of cosmology: Nestorian-Christian in Persia; Hindu-Greek in India; and Alexandrian Greek. Alexandria had

had its fabled library destroyed by Christian zealots in 389 but would long remain quite the most important Egyptian city. In due course, Hindu numeration became the Islamic norm; and would be received into Christendom as "Arabic". A "zero" symbol was the key to its efficient and aesthetic notation.

From 786 to 833, Greek texts were collected by Hārun ar-Rashid and his successor as the Baghdad Caliph. By 800, too, routine astronomic observation was under way within *dār al-Islām*. With this came the development of astrolabes in the eighth century AD and celestial globes by the ninth. Moreover, until the thirteenth AD/seventh AH quite a tradition was upheld of basic speculation, philosophic and cosmological.[19] Bukhara's ibn Sina (alias Avicenna) lived 980–1037 AD. Famous as a medical practitioner, he also kept astronomical records and edited a compendium on Ptolemy.

Avicenna always insisted he was a devout Moslem, his penchant for wine and women notwithstanding. However, he and certain contemporaries were strongly appraised of a rational obligation to seek progress through doubt, more so than anyone in Christendom appeared to be in his time.[20] As distinctive was his notion that the transcendent love of we humans for Allah affords linkage between the mundane world around us and His supreme intellect. Avicenna had defined positions on a number of other metaphysical propositions. He found astrology to be in practice unmanageable although, he felt, conceptually sound. He well understood that "its contemplative and symbolic side confirms closely to the basic spirit of Islam which is to realize that all multiplicity comes from Unity".[21] Much the same could apply to how, like Christian scholastics later, Avicenna was much taken by the Aristotelian doctrine of "universals".[22]

Averroës (Ibn Rushd) (1126–98) is the high medieval Islamic philosopher most often cited alongside Avicenna. Coming from Córdoba in *al Andalus*, he was long the more regarded among Christians. Broadly influential, too, was Ibn Khaldun (1332–1406). However, his scientific interests were more terrestrial. His best known work, *Muqaddimah*, dealt with environmentalism; and was, in fact, quite strong on climate determinism. Like Avicenna and other Islamic cosmologists, he was doubtful about astrology.[23]

An eventual decline of Islamic cosmology/philosophy cannot be just put down to a simple failure of adaptation. It relates more specifically to how *dār al-Islām* long encompassed no territorial heartland sufficiently cohesive and resourced to allow of due mobilization. According to hallowed principle, the one caliph should be the religious and political head of all Islam. But the interposed deserts and seas inhibited this.

Between AD 780 and 1171, there were always two or three Arab caliphates. At the turn of the Christian millennium, their respective locations were Córdoba, Cairo and Baghdad.

These individual polities also found it hard to engender firm statehood. Witness the impressment of infidel captives into military service. Nor was the Ottoman Empire to prove a helpful departure. All else apart, its overriding identity was Turkish not Arabic. Besides which, its capital from 1453 – Istanbul, formerly Constantinople – became a study in morbid obscurantism, a collective mind set typified by the rambling Topkapi (the Sultan's main palace) and, above all, by its harem. In the latter, hundreds of women and eunuchs lived lives of tedium relieved only by fear: "a minor breach of etiquette . . . resulted in instant death. In the seventeenth century, Sultan Abraham had 280 of his women sewn into weighted sacks and thrown in the Bosporous".[24] The outlook that allowed of this had also long engendered a deeply negative attitude to the adoption of gunpowder, watermills, windmills and printing. Hardly a good background for political and philosophical progress.

A Quest for Strategy

On various occasions this author has expressed his personal concern about the contemporary lack of cogent political thinking within the developing countries. The Moslem Arab world has by no means been an exception, this not surprising in the light of what has just been said.[25]

Gamal Abdel Nasser's *The Philosophy of the Revolution* was an anecdotal discourse that did much less than justice to the personality of its author, let alone guide Egypt towards democratic modernity. Otherwise, political development along the lower Nile has been by the Moslem Brotherhood, a movement founded in 1928.

Soon it adopted as its overarching motto the following: "Allah is our objective; the Qur'an is our constitution; the Prophet is our leader; *jihad* is our way; death for the sake of Allah is our wish". It has been committed to such elemental Qur'anic principles as addressing illiteracy, ill-health and dire poverty. However, it has lacked a discourse in depth on social transformation. Instead it had turned its focus more towards Arab international issues – notably Algeria and Palestine. The accent of its mainstream has lately been on non-violent remedies at home or abroad. But in 1954, Nasser repressed them, on a large scale and too brutally.[26]

A Brotherhood government democratically elected in the Arab Spring of 2012 was overthrown in 2013 by an Egyptian army anxious, in part, to retain its position of entrenched privilege within the body politic. An attempt to resist by unarmed demonstration was crushed ruthlessly. By early 2015, the states labelling the Brotherhood as "terrorist" included Bahrain, Egypt, Russia, Syria, Saudi Arabia and the United Arab Emirates.

The other Arab political movement of note historically is Ba'athism. It is a loose assemblage which developed in the Fertile Crescent, notably towards the end of World War Two. Inclined towards "socialism" (however defined) in its pursuit of a "cultural renaissance", it has been subject to various influences, French Marxism included. Its most prominent leadership figure, Michel Aflaq (a Christian "proud of Islam") was persuaded early on that secular socialism needs be tempered by spirituality. Repeatedly, the movement was to be riven by tension between Damascus and Baghdad.

The Shi'a

Still quite the most significant sacerdotal and political cleavage within Islam remains that between Sunni and Shi'a, a cleavage originating in early clashes between the family of 'Ali (cousin and son-in-law of the Prophet) and the Ummayads, the leading Quraysh clan. The key Shi'ite claim has been that 'Ali and next his grandson Husayn should have succeeded the Prophet as spiritual leaders. Instead, in AD 680, Husayn and his immediate followers were lured to Karbala in Mesopotamia and murdered. Today the Lebanon is the only Arab state that is demographically mainly Shi'ite. But Iran is as well.

Sufism

The Qur'an tells how the Prophet Mohammed received the word of Allah through mystic experience (*sūra* LIII, v. 1–18). A Sūfi ascetic-cum-mystical tendency proliferated in the centuries to follow, not least among the famed poets of Shi'ite Persia, alias Iran. Outstanding among them was Omar Khayyam (1048–1131), now recalled as a poet of world renown but, in his time, a peerless mathematician and court astronomer as well. The Sūfi aim is to draw nearer Allah through sheer love of Him, reducing thereby the void caused by our inability to comprehend Him through reason or describe Him prosaically.

The influence of sūfism has ramified. It has spearheaded interaction with other creeds, inputs from the New Testament and from Buddhism being especially evident. Sometimes it is forging links within the *umma*, the community of the faithful. At other times, it is pressing for social improvements. It does seem more open than other Islamic traditions to religious syncretism. How well it is surviving the current turmoil is another question.

Folk Memory

Still, History is important as a store of folk memory, not just as an explanation of the contemporary scene. By capturing Constantinople/Istanbul in 1453, Islam gained its biggest victory ever against Christendom. But 1492 was to be a milestone year in a much broader offensive by Christianity. Columbus in the West Indies set Renaissance Europe on the way to establishing a network of sea routes that would, among other things, encompass Islam. A reaction was predictable. The surprising aspect is that it took so long coming.

From 1763, Muhammad ibn Abd al-Wahhāb (*c.* AD 1703–91), a former Sūfi adept, waged tribal war to impress on Arabia and, hopefully, Indian Islam how the vanquishing of Satan depended on a return to the Qur'anic doctrinal purity of the early centuries, this as confirmed by *i'jmā,* i.e. consensus, be it popular or scholarly or priestly. Wahhabism thrived sufficiently to put Ibn Saūd on the throne of what thereby became the Kingdom of Saudi Arabia in 1932. By which time, oil-driven Western imperialism was inducing culture shocks in a region which was and still is much recognized as the spiritual heartland of Islam.

The "Islamic state" construct of 2015 is likewise reversionary, save that it or something like it has much wider aims geographically – a kind of Limited World War. By any standards, it is nearly as terrorist within *dār al-Islām* as Nazism was within Christendom. May one surmise that one factor underlying its exceptionally crude appeal is simmering anger over the Palestine question, not least because it has been a prime source of almost endless Arab defeats on the battlefield. What is desperately needed in any case is progress towards an equitable two state solution of this "Hundred Years' War". The centenary of the Balfour Declaration comes in 2017.

Still, it may well be that the major faiths, Islam included, will become more devolved as this century progresses. There have lately been indications of this in Islamic South-East Asia. Sūfism has shown signs of regional revival; and, in Malaysia, the Islam Hadhari movement has stressed the ethical merit of being both "cosmopolitan and Islamic". Much may depend on whether Indonesia can achieve lasting political stability in spite of the acute demographic imbalance between overcrowded Java and the outer islands. Much may depend, too, on a deeper engagement with modern scholarship, not least cosmology. This can likewise apply to Islam as a whole.

Scholarly Coverage

A short but noteworthy text was published in Riyadh, the capital of Saudi Arabia, in 2007. Its aim was to demonstrate how extensively the philosophy and findings of modern Science were anticipated in the Qur'an.[27] The binary character of so much of Creation was one especially salient theme. A number of its other contentions are debateable. Thus mountains are portrayed as pegs of stability rather than as the fulcra of seismic change they have so often turned out to be (p. 31). Nevertheless, the text as a whole can be said to have put well *a priori* a personal case for deism, religion whereby the Almighty does not intervene directly in the running of the cosmos.

One should be especially mindful of the richness of the Islamic astronomical tradition. During the recent Christian turn of millennium, it was estimated that 10,000 Islamic astronomical manuscripts and close to a thousand Islamic astronomical instruments were to be found across the world.[28] Many more such manuscripts, in particular, had been lost to posterity. Moreover, it is evident that, until 1500 AD or so, Islamic cosmological thought was interacting extensively with the Classical Greek, Persian and Indian schools. But afterwards this died down, considerably due to 'geopolitical factors'. However, it could be expected to figure in a general revival of Islamic thought within a cosmopolitan context.[29] But bedouin star lore has to be seen as another doomed tribal culture.[30] The lore about science (and especially cosmology) in the great civilizations of world history is still inspirational, collectively and individually. But the inspiration can only be transformational when and where the current science base is adequate. The average Islamic country today has nine scientists and engineers for every 1000 people; the global mean is 41. Only three Nobel prizes have so far been won by Muslims.[31]

(2) The Holy Land

Simply to cast in broad outline the often negative interaction between Islam, Judaism and Christianity could be to play down unduly the nodality of the Holy Land in terms of tensions rooted in religious differences. Here then is an endeavour to place Palestine four square in our consideration of (a) the development of relations as between what we too loosely term the desertic faiths and (b) what place religion may still have in what has started to look like an emergent planetary culture. It might be helpful to feed in a few impressions formed in-theatre, when working as a defence journalist (1966–73).

May this erstwhile early teenager of 1945 first recall how disturbing that year of victory was. A welter of data about the brutish in humankind came down the air waves hard upon Nazi Germany then Imperial Japan capitulating. Two sequences are locked in my mind. The one is of Hiroshima then Nagasaki after "the Bomb". The other concerns the Belsen concentration camp following its liberation by British troops, among them a sergeant my folks knew well. All bar one million of the seven million who died in such camps were Jewish. Zionism was thereby supercharged with anguish and resolve.

As the nineteenth century unfolded, "anti-Semitism" (i.e. anti-Jewishness) had revived across Europe yet again. It had done so against the background of a general upsurge of nationalism, this exacerbated by the long economic recession from 1873; and the sense of our Eurasian peninsula now being possessed of too much aggressiveness for horizons no longer expandable. As regards anti-Jewish sentiment specifically, the infamous *L'Affaire Dreyfus* broke within the French military in 1894. A decade earlier, following savage pogroms and anti-Judaic laws in Russia including Poland, Jewish emigration therefrom had begun *en masse*. By then, too, the emigration of European Jews to Palestine had significantly increased. "Next year in Jerusalem" was beginning to look less like a daydream. Zion was the Hebrew name for the hill upon which the city of David had been founded.

Debate got under way about a prescriptive agenda. In due course, the viewpoint of a charismatic Viennese journalist, Theodor Herzl (1860–94) prevailed. Shocked like so many by the framing of Alfred Dreyfus, a Jewish officer in the French army, for supposed treason in cahoots with Germany, Herzl became convinced the Jewish people must lay claim to their own state, a modern secular polity committed to justice and toleration. He concluded that this should be in Palestine, in Canaan. Save that statehood was not mentioned for fear of an Ottoman reaction, his thinking found expression in the Basle programme adopted by the first World Zionist Congress held, with Herzl as president, in 1897 in Basle.

The Balfour Declaration

Twenty years later, against the background of "the war to end all wars", this seminal Declaration was sent to Professor Chaim Weizmann, the then head of the British Admiralty's munitions laboratories, by Arthur Balfour, Foreign Secretary in David Lloyd George's wartime coalition. It confirmed that His Majesty's Government would back a Jewish "national home" being established in Palestine provided nothing "be done which may prejudice the civil and religious rights of existing non-Jewish communities in Palestine or the rights and political status enjoyed by Jews in any other country". Parcelling out the post-Ottoman Middle East, London and Paris had already agreed covertly that Britain should have the mandate for Palestine.

Balfour himself was at the centre of a "network of thinkers" resistant to the scientific naturalism of T. H. Huxley and his ilk; and no less determined to preserve a social and intellectual elite in the governance of Britain. Though he came to be labelled a Liberal Christian, his chief concern was the philosophical defence of theism. He would therefore have found hard to accept the notion that an NEO impact might randomly wipe us out.

He was closely linked with the Cambridge physicists who deployed the ether theory to underpin the notion that the physical world constitutes a unified whole, designed by God. He shared Oliver Lodge's interest in spiritualism; and served as President of the Society for Psychical Research in 1893–4.

In 1921, Prime Minister Lloyd George moved Winston Churchill to the Colonial Office. One task awaiting him there was implementation of the Balfour Declaration. He had himself been, since 1906 at least, "very well disposed towards Zionism".[32] It was a pitch consonant with his empathy with the Boers, alias the Afrikaners. In his Commons maiden speech, he had praised the Boer as a "curious combination of the squire and the peasant" and told how ashamed he had been to see such men ordered about "as if they were private soldiers" by young subalterns, a reaction occupation forces everywhere should still be alive to. Moreover, the Boers were a "People of the Book", the Old Testament. He found them possessed of industry, integrity, enterprise and dependability. The same applied to the Jews. Lloyd George thought likewise.[33]

Social Engineering

Churchill and the others concerned should have recognized throughout that the nub of the problem was preparing the Palestinians (through a major exercise in social engineering) for an accommodation any people, not least the British then or the Israelis today, would have found hard. One has in mind the intensive programmes the Americans introduced in Greece and Japan respectively post-1945, this in order to save each nationhood from further experience of the politics of extremism. Indeed, the Marshall Plan stemmed from very similar concerns.

Early in 1986, Shimon Peres, then the Prime Minister of Israel, urged that a quasi-Marshall Plan, led this time by Europe and Japan, be launched in the Middle East to offset recent falls in oil prices. He said he had lately spoken thus to European and Japanese officials: "If you want to guarantee the oil supply in the future you should invest not only in refineries and pipelines but in political infrastructure".[34]

Mutual interactive development could never have been effected in the Holy Land entirely spontaneously. All else apart, the Palestinian political culture lacked refinement after centuries of soporific yet repressive Ottoman rule. Local chiefs, imams and so on ran competing fiefdoms. There were urban elites not too antipathetic to some progressive change. Otherwise what passed for national ideology had a fascistic streak in it. Witness a memorandum presented to Colonial Secretary Winston Churchill in March 1920 by the Executive of the Haifa Congress of Palestine Arabs. One passage read: "The Jew is a Jew all the world over . . . He encourages wars when self-interest dictates and thus uses the armies of the nations to do his bidding".

Soon intensive debates developed throughout the Jewish diaspora about the merits or otherwise of Zionism in whatever mode. Two Leftist Jewish perspectives might here be especially considered, those of Hannah Arendt (1906–75) and Albert Einstein (1879–1955).

Arendt and Einstein

Hannah Arendt comes across as having been someone who regularly combined rock-solid courage with acute sensibility, nay spirituality. She still has much to say to our times. Thus she stood out, early on and very alone, against the manichean disposition to make "terror" a corral word which encompasses any forceful action by insurgents but nothing repressive their governmental adversaries do. She reminded us that an authoritarian government may operate a "terror" so ubiquitous that "often its victims are innocent even from the standpoint of the prosecutors".[35] Though she especially had the Soviet system in mind then, the point can well be taken more generally.

Einstein was a committed Zionist within the social democratic mainstream that led Israel to statehood in 1948. Regarding the Palestinians, he was looking by 1940 towards forging "an advantageous partnership which shall satisfy the needs of both nations". But in 1944 he felt constrained to warn that any such partnership can succeed only if both peoples conquer "that childhood complaint of narrow-minded nationalism imported from Europe".[36]

Had he been a shade younger, Albert Einstein might have served Israel (and therefore the world) bestirringly well as perhaps its second President. One does have to acknowledge, however, that his generosity could slither into naivety. Thus there was something rather callow about Einstein's extolling Gandhi as *the* greatest man of the twentieth century, this after his having himself to remind the latter pointedly that "non-violence" would be of little avail against Hitler.

Einstein was likewise too sanguine in imagining that, as a rule, the volunteer elites labouring to spread settlements into the desert would *ipso facto* be the people with the "humane and worthy" approach needed to "establish healthy relations with the Arabs which is the most important political task of Zionism". Surely the adaptive talents respectively required for these roles are too contrasting. And what about the social engineering needful on the Arab side?

The Jewish influx into Palestine during the interwar period soon began to register practical results – e.g. the building up of Tel Aviv, the development of the kibbutz movement, and a technocratic approach to water management. That the achievements were impressive is undeniable. Yet how sharply they contrasted with the historical Arab record tended to be exaggerated, not least in several scholarly works.

Take, for example, a 1944 commentary by Walter C. Lowdermilk, an Assistant Director of the US Soil Conservation Service who was to become Chairman of the UNESCO Commission on the Arid Zone. He wrote thus, "The decline of Palestine's land and of the people began with the first Arab invasion during the 7th century of our era. However, several centuries were to elapse before the country reached the stage of utter desolation. Although the nomads destroyed many cities and overran cultivated areas with their destructive herds, much of Palestine seems to have escaped the full effects of this first wave out of the desert. The country still was fertile and many of the native population were allowed to remain and continue their traditional ways of farming or carry on their former trades".[37] Evidently, this summation was somewhat contradictory. It was so because the author was (a) possessed of great expertise concerning agriculture at the arid margins and (b) a liberal Christian Zionist keen to see a national home in Palestine as a means of affording the Jewish people security while vitalizing anew the Holy Land via a Jewish–Arab partnership, its crowning triumph perhaps being a Jordan Valley Authority. But what of possible climate change, historically and in the future?

Worsening Aridity?

In 1966 an inverse link between agricultural bounty and the early impact of martial Islam was to be called into question by Rhys Carpenter, a classical archaeologist at Bryn Mawr. He pointed out how in nearby Anatolia, which the Islamic raiders never then reached, many churches were built in the fifth and sixth centuries. Yet come the seventh, no more were. Instead, existing ones fell into decay. Syria the raiders did occupy but as something akin to liberators. Yet here, too, church building died away.[38]

More conclusive, however, is the direct evidence of a regional climate trend towards worse aridity as medieval warming began to set in globally. In 1998, Arie Issar, a ranking Israeli hydrologist at the University of the Negev, went further than anybody before about such a switch in the Levant as a whole between 500 and 700.[39] Likewise, Amos Frumkin and colleagues at the Hebrew University, having measured drainage channels in bedrock in the northern Negev, concluded that *c.* AD 600 the water cycle got weaker there remarkably quickly.[40]

These qualifications illustrate all too well the simple truth that the Earth sciences are so often in need of multiple explanations that it is too easy to distort causation merely through omission, whether by accident or through intent. Something this means more specifically for us today is that the contemporary long round of global warming will likely further exacerbate the savagery today consuming the Middle East. This is because global warming will tend to push climate zones towards the poles. In particular, the Arabian desert will encroach on the Levant, Palestine included, more than ever before in recorded history.

Yet even today, the decision makers in international security affairs find it curiously hard to integrate global warming into their wider perspectives. At the time of the Balfour Declaration, prolonged climate adversity or, indeed, amelioration was on no official agendae. But the inspired Serbian physicist, Milutin Milankovitch, was by then working hard on what became his magisterial study of natural fluctuations in climate over multi-millennial timescales. He actually did quite a bit while on parole as a prisoner-of-war of Austria-Hungary. One can only hope that climate change will be allowed for appropriately as and when a two state or confederal solution to conflict is reached as between a more liberal Israel and a more liberal Palestine.

Active–Reactive

As of now, however, those concerned are sliding away from that consummation yet again. A fundamental reason has to be that each side has become imbued with a resilience so strong that it simply cannot see how tough its adversary has likewise been made by a burning sense of historical injustice. Correspondingly, there has been a basic disregard of the paramount principle of enlightened religion. Do not treat others worse than you would have them treat yourself.

In the aftermath of the 1948 war, the Israeli response to cross-border attacks was very generally one of minimal military action coupled with an invitation to peace talks without preconditions. But with Likud (i.e. the Israeli Revisionist Right) coming to power in the Knesset in 1977, the balance shifted more towards heavy retaliation. An especially terroristic case in point was Ariel Sharon's connivance in the massacre by Lebanese Christian zealots of 1700 of the inhabitants of the Palestinian refugee camps of Sabra and Shatilla, this at the climax of the Israeli invasion of the Lebanon in 1982. The parliamentary ascension of Likud was partly, of course, action–reaction to Palestinian insurgency. But it also owed a lot to an influx of harassed Russian Jews taking place.

Begin was to remain Israel's Prime Minister until 1983. I had interviewed him for forty minutes in 1972. Alas we had talked past one another the whole time. He seemed to me to be consumed by a kind of dark mysticism that barely connected with the world outside. He was for sure ridden with contradictions. Shimon Peres said of him that "He hated the British but spent most of his time listening to the world service of the BBC". The interview in question rates as the least substantial yet most revealing I have ever been given, whether in journalism or academia.

In the mandate decades of the Zionist experiment, sentiment had waxed strong that its main thrust should be towards social democracy. Much encouragement for this was forthcoming from the Jewish diaspora and from a broadening swathe of progressive gentile opinion. Post-1967, however, it came to seem to some of us in Britain at least rather as if many people in politics, the media, the churches, academia, the public service and so on had taken a vow of silence on the whole subject of Israel and the newly occupied territories.[41] Within Britain's labour movement, Apartheid in South Africa was debated one-sidedly and endlessly, comparable ethnic tensions in the Holy Land hardly at all. No doubt this selective abstention owed something to a desire not to revive customary "anti-Semitism" – meaning, of course, anti-

Jewishness. It will also have been an attempt to assuage well merited guilt feelings about the significant part progressive opinion had played in the appeasement of fascism. Hitler would have been so easy to stop before 1938, given a modicum of military preparedness and resolve all round.

A preview of perhaps changing times came in 2014 when Britain's Parliamentary Labour Party voted solidly in favour of a Palestinian state being established alongside Israel. Meanwhile, American opinion as articulated in its up-market journals was becoming more even-handed in its coverage of the Holy Land imbroglio.[42]

An Assize of Arms

One might say that in most colonial situations (post-1815, say) armed violence could accelerate or retard the arrival of independence though not necessarily alter its character much. In the Holy Land, the relationship has been, and no doubt will be, different. As much can be read into why, in the sombre victory autumn of 1945, Hannah Arendt had warned that "If Zionists persevere their sectarian ideology and continue with their short-sighted realism, they will have forfeited even the small chances that small peoples still have in this none-too-beautiful world of ours".[43]

Not that the sectarianism she perceived could ever be described as simply unilateral. On the contrary, Zionism was split between a social democratic mainstream by then led by David Ben Gurion (1886–1973) and the Revisionists lately under Vladimir Jabotinsky (1880–1940). The latter was a multilinguist from Odessa and a gifted orator. Careless of Arab rights, he and his colleagues aspired to create a Jewish state extended to biblical borders beyond the Jordan. Ben Gurion had come up through Polish trade unionism.

In 1931, a Revisionist guerrilla group, the Irgun Tzvai Leumi, broke with a Haganah by then established as the main Jewish defence force. The Irgun's most infamous action was the 1948 massacre of maybe 250 peaceable Arab villagers at Deir Yassin, a village in the Judaean hills. Much gratuitous mutilation was involved. This act of terrorism was condemned by social democratic Zionism as well as by the world at large.

Working as a *New Statesman* journalist in September 1967, I met on the East Bank of the Kingdom of Jordan not a few of the 90,000 refugees who had fled across the river Jordan because of the recent June War. Before leaving London, I had a briefing at the Israeli Embassy. I was told, in a spirit of constructive goodwill, that I would hear "countless atrocity stories in Jordan" and that "we categorically deny them all in advance".

On location, I talked to a fair number of the refugees. Just a few did tell of advancing Israeli troops pressurizing them verbally to "Go to King Hussein". There was, as I recall, just the one allegation of an air attack on a military ambulance convoy. But overwhelmingly the reason given for exiting was the abiding folk memory of Deir Yassin.

Before 1967, however, Israel's War of Independence had taken place, as had the "Suez Crisis". As trustee for the Palestinian mandate, Britain's declared aim in 1945 had been to preserve a communal balance sufficiently well to allow of the emergence of a binational state. From 1936, Arab insurgency had been a big security problem. But as the concentration camps were liberated in 1944–5, many Zionists came to find British compromise intolerable. Jewish guerrillas were to kill 150 identified adversaries during 1947, some 30 of them Jews seen as too imbued with the spirit of compromise.

In mid-1947, Britain formally referred the Palestine question to the UN. That November, the General Assembly endorsed a partition plan. It provided for the joint control of Jerusalem. It also divided the rest of the territory in such a way that neither community could lay claim to a contiguous bloc. Introduced several years earlier, it might well have been the basis for a binational confederation, one in which each member might have its own security forces. Haganah was prepared to give the plan a try. But the slide into resolution by force of arms had probably gone too far. At all events, Britain announced it would withdraw from Palestine on 15 May 1948.

At the start of the ensuing war, the several Arab armies involved numbered 90,000 men with 200 tanks. Against them stood Haganah's 10,000 riflemen and several tanks. To which may be added the several thousand Revisionist gunmen they collaborated with contingently – e.g. in the forceful expulsions from Jaffa to Gaza and elsewhere.

Yet Haganah was to be almost comprehensively successful militarily. Two noteworthy exceptions were (a) Jordan's British-led Arab Legion retaining the old city of Jerusalem and (b) Egyptian troops under Gamal Abdel Nasser holding out well in the Sinai. Otherwise the Arab armies showed a desperate lack of motivation throughout their command structures.

Armistice agreements were signed all round in 1949. By the start of that year, over 80 per cent of the 850,000 Palestinian Arabs had left their homes whereas 750,000 Jews were now ensconced within the mandate boundaries as against 550,000 in 1945. A debate has raged ever since as to how far the Palestinian exodus was enforced by their

Zionist adversaries. That there were ruthless expulsions is evidenced in the Jaffa example already cited. And in Haifa the Palestinian residents were given but the briefest of intervals to reach an understanding before eviction commenced.

More generally, though, there may well have been substance in the longstanding Zionist contention that the Palestinians exited because broadcasts from Damascus encouraged them to step aside for the short time needed to secure victory. But it also seems true to say that many who sought in due course to return found their way blocked by the Israeli military. That outcome can hardly be described as remarkable. But it is important that it regularly enters the record. This whole dispute has been plagued throughout by information bases partial in both senses.

1956 and 1967

The next warlike crisis, "Suez" in 1956, proved to be the most convoluted of the series. My claim would be that a summation sufficient for current purposes was proffered in *The Geography of Human Conflict*.[44] Suffice now to identify what one sees as three salient weaknesses in the case made by the critics of the Anglo-French expedition, those in Britain for sure. They said little, then or subsequently, about the manner in which the Eisenhower administration in Washington disengaged from its provisional commitment to finance the construction of the Aswan High Dam. They said nothing, then or subsequently, about Israel's involvement although suspicions were immediately aroused about Jerusalem's collusion with Paris and London, suspicions that were to be confirmed by early 1957.

Surely, though, the crux was that Israel's air arm was by no means as strong then as it would be come 1967, the date of the next big Arab–Israel encounter. Therefore the crippling of Egypt's air force through the RAF's bombing of her airfields was what ensured the Israelis could complete within five days their advance across the Sinai. Had this to take much longer, it might well have collapsed instead. As things were, many of the 5000 vehicles involved were so unsuited to the rough desert tracks that 2000 reportedly broke down beyond immediate repair.[45]

So what alternative did the critics of the Suez imbroglio have in mind? Could they endorse Washington's deciding to withdraw support from the High Dam project? Were they enamoured of how brusquely that decision was communicated? Had they been pressing for incremental progress on the Palestine question? What military contingency plans should the West have had in place beforehand? Would any of the critics have regarded the military defeat of Israel as a stabilizing outcome? We never had answers. We never shall.

The proximate cause of the 1967 Arab–Israel war does appear to have been Moscow feeding its acolytes in the Middle East spurious intelligence about supposed Israeli preparations for an offensive against Syria. Getting wind of this, the Israelis invited the Soviets to provide some shreds of evidence. No matter. Egypt under President Nasser decided to form a containment alliance with Jordan as its centrepiece. King Hussein agreed.

Oxford's leading specialist in recent years on the Israel–Palestine dispute has been Professor Emeritus Avi Shlaim, very much an Israeli peacenik. His verdict was to be that "Hussein made the mistake of his life by jumping on Nasser's bandwagon. The price he paid was the loss of half his kingdom, including the jewel in the crown, the old city of Jerusalem".[46] Effectively, the loss was incurred the first morning of that June war, with the Arab air arms being immobilized on the ground in a devastatingly professional Israeli pre-emptive air strike. After that, Israel's total victory could only be a matter of a week or so.

In Amman that September, one was to be told on very good authority that the Jordan Arab Army's mind set had been that it would fight under conditions of air parity. It sounded like but another instance of warriors in light blue simply being discounted by those in khaki. Yet the battle for Jerusalem apart, it never made sense to think that tactical aviation could thus be written out of this assize of arms. Air superiority secured over open landscapes has very consistently yielded a decisive advantage overall, this past century.

At all events, the small Jordanian air arm, however proficient and dedicated, could never have been remotely a match for the Israeli. Nor could Egypt's air arm have expected to extend protection over the Hashemite kingdom. Syria, too, would have found providing this cover infeasible. Moreover, there was for Damascus an underlying issue as well. It had lately been subjecting Amman to a splenetic propaganda offensive. The king himself had been described as a "jackal" whose mortal remains ought to be "nailed to the door of the British Embassy" as soon as possible.

The comprehensive defeat of the regular Arab armies in the June War was to swing Palestinian opinion rather more towards *fedayeen* insurgency. Accordingly, the bazaar area leading down towards the Roman amphitheatre came to be parcellated by road blocks controlled by the Palestinian groups, the police and the army in various permutations. An incongruently large Soviet embassy was clearly master-minding a flow of military supplies to the insurgents. However, insurgency incursions into the newly occupied West Bank were subject to heavy losses. Therefore by 1970, the *fedayeen* were turning to aerial highjacking as an alternative. So that September King Hussein resolved to end the anomalous situation in Amman. He ordered in a task force headed by bedouin armour. The Palestinians fought hard and unitedly for a week or so but were overcome. One may surmise that they had failed primarily because Moscow had supplied no antitank weapons – i.e. mines or rocket launchers. Doing so would have conceded to the *fedayeen* too much autonomy.[47]

The Impact of Allon

Bearing in mind the folk memory of Deir Yassin, it would surely have been better if the Israeli government had put the deadline for the return of the new wave of refugees created by the June War of 1967 back to a date significantly later than 30 September of that year. This would have given all concerned time to take stock coolly, this including perhaps some dialogue with their adversaries.

The view that this could have been helpful did, in my observation, receive notable support from the Israeli Embassy in London. However, the instinct that prevailed in Tel Aviv/Jerusalem was that it was best swiftly and firmly to take the helm, to show who was now in charge. Acting in the same spirit, a senior government minister, Yigal Allon, enunciated as early as July 1967 a plan for the future of the occupied West Bank. Within weeks, work had started on a string of small settlements down the Jordan valley. Their locations were to be so chosen as to (a) provide early warning, (b) block mechanized incursions and (c) allow of counter-attack. Their operational utility could be questioned. But the symbolism was ominous. So was the proposed absorption completely into Israel of an expanded Jerusalem. No matter that Arab national aspirations did receive, at the same time, routine recognition.[48]

A plan as characterized by loose ends as was Allon's lent itself either to disregard or else to unduly free interpretation. Apprehensive of this, David Ben-Gurion earnestly opposed in his last few years all settlement in the newly occupied territories. That he could no longer prevail was all the more regrettable in that, at middle management level, the Israelis were pursuing some enlightened social policies in these territories – a freer press, health care, agricultural development . . .

Talking at length to Allon along with a ranking Jewish author of leftist inclination in the autumn of 1967, one found the former amazingly impassive. He remained so even when my colleague advised him that regaining the Wailing Wall was a historic opportunity to break out of the Zionist *leitmotif*, perhaps through demolition. Nor did Allon evince the slightest hubris about the recent triumph. His dourly implacable pitch throughout was that "we have had enough" of Arab rejectionism as expressed in incursions from Gaza and bombardments from the Golan.

But hubris soon became evident more generally, this tending to favour the political Right. During a 1972 visit as a guest of the Israeli government, I suggested to a senior retired member of the Israeli defence planning cadre that it might be simplest to go for a binational state. He said "I don't think our own people are capable of it". He was acutely concerned about the whole outlook.

In the autumn of 1973, Egypt now under Anwar Sadat launched a full-scale offensive across the Suez Canal, a stratagem which brought Israel closer to defeat than at any time since her inception. The key to its qualified success was two-fold. Arab governance had apparently been made less widely open to Israeli espionage than it had usually been before. In addition, along the canal itself a sophisticated Egyptian deception plan had been implemented with consummate competence. Therefore Israeli apprehensions that something was afoot did not firm up sufficiently to allow of the Americans being told that another preemptive strike was imperative. Eventually, Henry Kissinger, then US Secretary of State, would make sure Israel was supplied with enough advanced American weapons to turn things round. Yet he failed to follow this up with a Middle East peace offensive.

Instead, President Sadat and Prime Minister Begin were to sign a separate peace treaty in 1979. Since when, the Egyptian army has turned inwards to protect its own vested interests, as intimated above, though also to shield the body politic against the Islamic militancy that effected Sadat's assassination in 1981. Meanwhile the situation in the Holy Land itself was continuing to give all too much ammunition to those the world over who say religious commitment is liable to lead to war.

Action–reaction continues to feed extremism on both sides. In particular, Israel's entitlement to regional acceptance through formal recognition has been badly compromised by Palestinian bitterness over, in particular, the illegal Jewish settlements together with the countless check points and, indeed, roads for the use of those settlers only. Once finally entrenched, apartheid in the Holy Land would be a disaster for both the peoples involved. Ultimately, it would be, too, for the planet as a whole. A planetary survival strategy cannot be progressed towards without genuine peace in the Middle East. And that peace cannot be forthcoming without an equitable and stable resolution of the Holy Land question.

It was salutary to find Ruth Dayan, as the elegant 95-year old widow of Moshe, warning all concerned late in 2011 that "Zionism has run its course".[49] In other words, Israel needs a new *raison d'être*. Her opinion would assuredly be endorsed by most of the greatest Zionists past: Hannah Arendt, David Ben Gurion, Albert Einstein, Yitshak Rabin . . . Probably, too, Moshe Dayan. He said just after the 1967 victory that the Palestinians needed 50 years to catch up. So a big peace drive to mark the centenary of the Balfour Declaration? And this a vital step towards rebuilding, across the war-torn Middle East, the democratic Arab Spring? And that a precursor of wider planetary progress towards enduring peace?

6

The Thermonuclear Five

However many games are devised to predicate a satisfactory outcome for active defence against incoming Near Earth Objects (NEOs), they will stand no chance of so doing unless a suitable geopolitical context can be looked towards. What this must avoid, first and foremost, is any development liable to tip towards unitary world government at some early stage. A general case for avoiding that outcome was put in *The Bounds of Liberalism*.[1] May one now visit it anew with the anti-NEO requirement especially in mind.

Uneven economic growth, registered or prospective, will be a big complication in world affairs well beyond the next turn of century. Relevant transnational comparisons were regularly being made ahead of the recent world recession. China's income per capita in 2050 was usually put at less than half the USA's still. So was Brazil's. Similarly, India's would yet be well under a third; and Indonesia's less than a fifth. Moreover, quite a few states would remain below that spread. Hardly a sound basis for a general merging of sovereignties, notwithstanding notable convergence in certain cultural aspects – e.g. youth lifestyles; science and especially medicine; and, incipiently perhaps, religious pan-ecumenism.

One must bear in mind throughout, too, how the manifold communications revolution so strongly under way will oblige all polities to become more open, more participatory or else a sight less so. Maybe the political geography of our planet will, in due course, become more sharply divided, a while at least, between those regimes that have gone further down the authoritarian route and those that have turned more liberal. Such a dichotomy could be enough in itself to preclude unitary though open governance.

Also to bear in mind is how ecological disjunctions across our planet further undermine hopes of a functional world democracy. Many such glitches look more acute now than seemed the case just a few decades ago. This may apply not least so far as the high seas are concerned.

Functional World Government

Then there is military rivalry to consider, not least in the strategic nuclear dimension. The British philosopher-cum-world historian, Arnold Toynbee (1889–1975) was Director of Studies at the Royal Institute of International Affairs, London (alias "Chatham House"), 1918–39. Writing in 1971,[2] he presumed "we all *knew* . . . how unstable and insecure" (p. 66), my sceptical italics, the nuclear balance of terror was – this regardless, so he thought, of its actual configuration. He inferred that this alone must ultimately involve sovereignty being surrendered to a unitary world government "likely to be imposed on the majority by some ideology or religion" (p. 113). He thereby joined the serried ranks of those whom some of us feel are too absolute in their nuclear apprehensions. But this does not mean his and other such concerns have been entirely without foundation.

Writing in 1949,[3] Bertrand Russell was finding it "often argued that an international government would be oppressive" and did "not deny that this might be the case at any rate for a time" (p. 183). But the average size of states had been growing. Now the "only two wholly capable of independent self-direction are the USA and the USSR. All that is necessary to save mankind from disaster is the step from two independent states to one – not by war, which would bring disaster, but by agreement" (p. 186). It will seem to many of us that this assessment rested on an optimistic interpretation of what should be meant by "all" in this context. Could such a binational agreement be reached? And would it actually work?

The Permanent Members

It could be well worth our while building to an extent on how the five permanent members of the UN Security Council (i.e. Britain, China, France, Russia and the USA) happen, more or less fortuitously, to be the only countries which have as yet proved able to develop not just nuclear fission warheads but also nuclear fusion alias thermonuclear ones or "hydrogen bombs". There is a fairly well defined gap in the spread of nuclear explosion strength between the most powerful of the fission warheads and the least powerful of the fusion ones. The former are in the several hundred kiloton range and the latter near to a megaton. Doctrine needs also embrace responses to chemical and biological threats.

All the members of the thermonuclear quintet use function built submarines as platforms on which to locate at least a proportion of the relevant missilry. Nor can it be denied that even a largish submarine, when fully concealed beneath the sea surface, has been, thus far, impossible to locate across anything like oceanic distances. Submarines submerged will give themselves away briefly when using a periscope or wireless or if firing a missile. None the less, a submarine force bearing thermonuclear strategic missiles has so far been a very stable form of deterrence. Pack ice covering polar seas may strongly secure their concealment. However, this security of obscurity is liable gradually to become less absolute overall. The seas are becoming more transparent, due not least to Russian innovation.[4]

Minimal Deterrence

Britain's Royal Navy has long maintained that, in order to ensure there always being at least one such submarine on station, it is necessary to have in commission at least four. The other three may respectively be (a) *en route* to or from a given patrol area, (b) in the home port for routine maintenance and crew recuperation or (c) undergoing a refit. The port facilities themselves will never be entirely safe from attack, nuclear or non-nuclear.

Thermonuclear Proliferation?

Other states are, of course, continually learning more about nuclear fusion as well as fission. But none, not even North Korea, seriously aspires to apply fusion knowledge unilaterally to military ends or, indeed, civil ones. Nor may it be at all likely that, if any other statehoods did resolve unilaterally to go militarily thermonuclear, they could achieve this independently in under twenty years.

If there is a country which might be able to come close technologically to realizing such an aim, it has to be Japan or just possibly South Korea. The former has been accorded a special status within the International Thermonuclear Experimental Reactor (ITER),[5] this by dint of her own progress to date in the civil thermonuclear field, progress extremely well exemplified by the facilities at Naka in central Honshu. She would have hosted ITER had not America's predilection for her doing so not been overridden by a European lobby. At present, Tokyo evinces no aspirations for military nuclear/thermonuclear autonomy.

A Thermonuclear Quintet

It might well be acceptable for a planetary drive for protection against NEOs and also terrestrial threats of comparable character and magnitude to be led by the existing thermonuclear quintet, maintaining as between themselves a rough-and-ready balance of power all round. Better that than no coordination of the crucial early steps in the creation of an anti-NEO screen. Far better that than would be the imposition of the required construct by a single superpower. One would need, in any case, a great deal of ancillary support and much collateral goodwill.

Hopefully, multilateral development initiated along these lines could evolve into something of an alternative to terrestrial warfare as a source of purpose and identity. It might move on further and provide a setting for a substantial planetary peacekeeping force – a thorough recasting of the contingency provisions the UN has lately been considering. Indeed, it might progressively pave the way for a diversity of international groupings, regional and otherwise. And these could lead to the emergence, this next century or so, of a loosely pluralist global regime. In the light of what has been said above about economic disparities, better "loosely" than "tightly".

European Adaptation

What must be conceded, however, is that a UN departure of the kind here being outlined might never take off at all so long as two-fifths of the voting strength of the Security Council's permanent membership remains Western European – i.e. France and Britain. Nor may a better answer be found by expanding the number of permanent member states. Attempting to go down that route for more general reasons has lately proved to be unproductive and divisive.

An evident alternative could be to have France and Britain work towards alternating or otherwise sharing just a single permanent seat on the Council instead of the present one apiece. Not merely would this correct, more or less, European over-representation. It would also underline the principle that London and Paris were there to represent a wider European perspective, EU and otherwise.[6] Should Britain's exit from the EU prove permanent, such collaboration need not be excluded. On the contrary, it might be a good alternative, from more or less all points of view.

Naval Integration?

Closely related to this possibility is the further one of integrating operationally the British and French strategic submarine forces. There was a flurry of interest in this particular option around 1970, the integration looked for then being in command structures, not the mixed manning of individual vessels. Such integration might allow of some financial savings but the primary consideration would be alliance solidarity. Edward Heath, as Britain's Prime Minister from 1970 to 1974, was actively interested in such integration possibilities, being himself a thoroughly committed "European". The issue may come up again before too long. It need not be seen as incompatible with Britain's leaving the EU. But it would require both countries to remain permanently within NATO as well as in the Security Council.

Currently, the Royal Navy is planning to concentrate in the Firth of Clyde area of Scotland the shore support for all its submarines, not just the strategic nuclear ones. But what will be the situation if, although Britain as a whole has voted in the referendum to leave the EU, Scotland prefers to stay in? How much sentiment is left in favour of the "auld alliance" between Edinburgh and Paris, a feeling of fellowship traceable to early modern times? Might an independent Scotland still in NATO lend logistic support to an Anglo-French deterrent force?[7]

Moscow and Beijing

A terribly ruptured Middle East is currently the most critical test case of what progress the thermonuclear five (or should one be saying "four"?) can make, in the near future, towards constructive togetherness. On 28 September 2015, Vladimir Putin addressed the UN General Assembly. He said the West had made a bad error in not working with Bashar al-Assad to defeat the "Islamic state" in Syria and Iraq – basically "the Fertile Crescent". In an oblique response, the Obama administration intimated that Assad might, after all, have a role to play in the transition to a new Syria yet to be agreed on.

Meanwhile, Russian air strikes were getting under way against "terrorists" opposed to Assad. The sinister "Islamic state" was flagged as the prime target. Duly, some informed Western commentators were welcoming, altogether too uncritically, these apparent signs that Moscow was coming on board with the West's intervention.[8] Most of the strikes Moscow then authorized against targets in Syria were specifically against rebels other than the "Islamic state". Should we have talked of a Limited World War against the influence of the West? And is that disposition still latent?

Old Believers?

In the seventeenth century, those conservatives in the Orthodox Church and elsewhere who were anxious to ensure Czarist Russia never experienced what might pass for a belated Renaissance dubbed themselves "Old Believers". From 2015, Vladimir Putin, a former Communist apparatchik turned thriving plutocrat, harshly confirmed his position as the Great Russian czar of a new generation of Old Believers.

Contrariwise, President Xi Jinping of China paid an official visit to the USA, September 22 to 25. He seemed to have made some real progress with persuading Barack Obama that "our two great nations, if we work together, have an unmatched ability to shape the course of the century ahead."[9] And even during that brief span, the world's top two personalities as of then specifically agreed not to conduct or support cybertheft against each other. They also looked towards a jointly progressive approach at the world climate conference due that November, a togetherness then carried forward convincingly.

Moreover, the two pledged themselves to do all they could to enhance strategic mutual trust and "jointly face global challenges". This would include *inter alia* China doing more in the way of well targeted economic aid. Also, Beijing was considering a contribution of 8000 personnel to a UN stand-by peacekeeping force being mooted. So was she emerging as tacitly a victor in the Syrian civil war, her operational non-involvement notwithstanding? Or is Russia still in pole position instead? And what difference might President Trump make early on? What Action and Reaction?

Thermonuclear No First Use?

Membership of what may informally be called a thermonuclear club brings with it both advantages and obligations. It does so because, as is corroborated by the chemistry and physics of every star so far checked out, there is, in principle, no upper limit to the scale of a thermonuclear reaction.[10] Through the climax of the Cold War, thousands of nuclear warheads were operationally deployed within the two major alliances, NATO and the Warsaw Pact. Quite a few of those borne on strategic missiles, strategic bombers and strike aircraft were fusion warheads in the low megaton range.

The NATO fission warheads were issued to theatre war forces. Many will have had rather less explosive power than either of the bombs dropped on Japan. Indeed, the smaller of them had fission yields down to a quarter of a kiloton. These would have been considerably for use as mines or else anti-aircraft warheads, air-to-air or surface-to-air.

Neither NATO nor the former Warsaw Pact (nor any other command to my knowledge) has ever as yet sought to distinguish in theatre conflict between the principle of no thermonuclear first use and that of no first use either of nuclear fission warheads. Yet the former precept could be deemed superior for deterrence but also for arms control. If it needed a rider, this might have to be that it could well be desirable to keep the thermonuclear first use option open also against an adversary state evidently accessing biological weapons.

Correspondingly, there is likely to be much discussion about whether we should use fusion or fission or neither to destroy asteroids and/or comets. Quite a lot of anti-NEO specialist opinion seems already in favour of making ready to use whichever is operationally needful. One therefore feels apprehensive that this issue will tend to wax too prominent within the anti-NEO debate. So much else will also need deciding.

7

Skyward Precursors

(1) Muscovite Adjustment

Some grounds still exist for cautious optimism apropos the next several decades. One helpful circumstance may indeed be that the permanent members of the UN Security Council are likely for some time to remain the only countries possessed of thermonuclear military status.

However, two members of this quintet – Russia and China – may have too often felt excluded, felt that too much of the Council's take on world affairs since 1945 has emanated from the Atlantic trio, above all the USA. Moscow and Beijing have respectively sought to correct this imbalance, in part through the influence attaching to up-front science. Each has an inspirational scientific legacy, not least in the sky sciences.

Peter the Great (1672–1725) set the scene, as Czar from 1682, for quite a strong development of natural science in modern Russia – most notably, albeit well after his time, the periodic table in chemistry. In 1703, he had founded St Petersburg to replace Moscow as the capital city. This it did from 1712 to 1918. Peter sought thus to create a Baltic "window on Europe".

Correspondingly, Czar Peter prepared for the establishment (in 1726) of an Academy of Science to spearhead scientific developments nationally. This came in the wake of such institutions being established in Western Europe. Thus in 1662 Britain's Royal Society had been formed. Four years later the Paris Academy of Sciences had held its first meeting. Russian science would still lag but no longer as comprehensively.

Bolshevik Advent

Quite striking in the century lately past was the virtual concurrence, in 1917, of Albert Einstein's General Theory of Relativity and the

Bolshevik revolution. Come 1945, Stalin had concocted a formulation which said that Soviet scientists were to make full use of Einstein's "objective" findings yet ignore the "bourgeois idealism" the Soviet dictator averred suffused them. However, in the pristine early days of the revolution, scores of thousands of literati across Soviet Russia had become involved in a Protelkult (proletarian culture) movement.[1] Led by Konstanin Tsiolkovsky (1857–1935), see below, many Protelkult members sought to hitch their futurist and millenarian aspirations to the stars, this via a credo of Cosmism, alias planetism.

Alas, this "flat romanticism" was scorned by Leon Trotsky:

> "We have conquered all Russia recently and now we are going on towards world revolution. But are we to stop at the boundaries of *planetism*? Let us put the proletarian hoop on the barrel of the universe. . . . We'll cover it all with our hat."

"Cosmism seems or may seem extremely bold, vigorous and proletarian. But in reality", Trotsky continued, "Cosmism contains the suggestion of very nearly deserting the complex and difficult problems of art on Earth so as to escape into the interstellar spheres".[2] Stalin, too, came to view Cosmism as retrogressive.

Konstanin Tsiolkovsky

In 1923, the year before Lenin's death, the Protelkult was abolished. But Konstantin Tsiolkovsky continued imperturbably to advance within the USSR the science of aeronautics and of prospective interplanetary travel, twin themes he had turned to in childhood to offset his early and near total deafness. In 1894–6, he had done paper studies on monoplane design as well as interplanetary rocket travel. Then, in the 1920s, encouraged by the Bolshevik regime, he was to do conceptual work on such heady prospects as multistage rockets, jet planes and air cushion vehicles. His work was soon to enjoy international recognition, Soviet impediments notwithstanding.[3]

Maybe it is appropriate to mention also the cosmic philosophy prominent within Tsiolkovsky's range of interests. He intuitively felt "There is no substance which cannot take the form of a living being. The simplest being is the atom. Therefore the whole universe is alive and there is nothing in it but life."

He addressed specifically what we have come to know as Enrico Fermi's paradox. If the rest of the universe is full of life, advanced and elementary, why is it that none seems ever to have visited our Earth?

In essence, Tsiolkovsky's own rather convoluted answer was that we ourselves have not been ready for it: "We are brothers but we kill each other, start wars and treat animals brutally. How would we treat absolute strangers? . . . Can we really have relationships with dogs and monkeys? In the same manner, higher beings are not able to communicate with us for the present?"[4] But how would they or we have known?

Overall, the Russian authorities, Czarist then Communist, constrained considerably the publication abroad of works by Tsiolkovsky and certain other astrophysicists. His 1903 study of proposed Space exploration via reaction-propulsion (i.e. the rocket principle) was banned by the political police. However, he did receive a Soviet state pension from 1921.

None of his writings were to be published in English until 1960. Nevertheless, Hermann Oberth, a German rocket pioneer, took a keen interest in his disquisitions on rocket-driven Space travel and colonization.

Kremlin Aspirations

Visionaries of Space travel subtly held out the prospect that even scientific materialists will, in due course, make it to heaven. Moreover long-range military rockets might (especially if complete with atomic warheads) considerably offset, at long last, Moscow's inadequate access to ice-free open sea.

Meantime the development of the Soviet North struck a chord with the puritan ethic that suffused Soviet Marxism. Not least remarked in this regard was the Northern Sea Route to the Pacific, extolled the more in the aftermath of Russia's crushing naval defeat by Japan at Tsushima in the war of 1904–5. The said reverse had highlighted for St Petersburg a contingent need for faster, smoother and more adaptive marine deployments to or from the Pacific theatre. However, come World War Two, only one part in 40 of American lend-lease aid to the USSR came westwards via this Arctic seaway.[5]

Towards the end of 1935, Stalin had begun to wax obsessive awhile about building the world's biggest surface navy. From 1938 several "Super Dreadnought" capital ships appear to have been laid down. But none was ever completed, given the exigencies of the Great Patriotic War.[6] Post-1945 the USSR made submarines the bedrock of its naval strength.

After 1945, too, the USSR developed nuclear bombs, fission then fusion, sooner than Western analysts expected; and complemented these gains with a civil-cum-military rocket programme. As is much recalled, her Sputniks 1 and 2 became in 1957 the world's first orbital Spacecraft; and in 1961 her Yuri Gagarin became the first human in orbit. As Nikita Khrushchev was to put it, "All over the world, our satellites go 'bleep, bleep' except over Washington where they go 'ha, ha'."

However, this still stands as the Muscovite high noon. The Russians have yet to land anyone on an extra-terrestrial body. Moreover, their attempt to outdo Neil Armstrong's lunar alighting by concurrently projecting an unmanned surveillance probe onto the Moon's surface was a fiasco. Subsequently, Moscow gave top Spaceflight priority to testing and extending human endurance in orbit. After the fall of the Soviet regime, she became a participant in the American-led International Space Station. To what avail is unclear.

It can be instructive to regard cosmic threats of destruction and death as an extension of those from terrestrial sources. A most basic difference between the two though is that we (meaning our planetary community at large) now have well over half a century's experience of managing strategic protection against man-made thermonuclear attack. What this has shown us, above all, is how easy it can be to make compromising errors in respect of lay-out, either institutional or material, not to mention the operational management of what might become a very fast breaking warlike crisis.

Missile Gaps

The build-up to, and the events during, the Cuba missile crisis of 1962 illustrated all this only too well. In 1955, American radar located in Turkey had detected a Soviet long-range rocket being test-fired. Then in 1957, the Sputnik launchings connoted the USSR's possessing surface-launched rockets sufficiently large and accurate for "intercontinental" (meaning, as per usual, inter-hemispherical) military nuclear engagements.

Apprehension expressed on the American campaign trail of an apparently imminent "missile gap" sufficed to clinch the desperately narrow victory Senator John F. Kennedy secured over Vice-President Richard Nixon in the 1960 Presidential. Nor was it just within the USA that the Sputniks were read as portentous. "The East Wind", said Mao Tse-tung, was "prevailing over the West Wind". Duly, the *Peoples' Daily*, mouthpiece of the Beijing Maoist establishment, advised its readers that the ascendancy of the "anti-imperialist bloc" had reached "unprecedented heights".

Soviet Misjudgement

The impatience of a mystified China with new-found Soviet caution increased during the Taiwan Straits and Lebanon summer crises of 1958. Indeed, the situation can be said to have marked the commencement of the great Sino-Soviet split. One could surmise further that the quiescence of Communist Eastern Europe between the invasion of Hungary in 1956 and that of Czechoslovakia in 1968 owed something to the Sputniks appearing to betoken an overriding superiority. The "missile gap" legend was so widely believed that it acquired a while a certain reality.

As to how the situation then unfolded, the Soviet planners were to prove altogether too preoccupied with Western Europe as either a potential hostage or else as a putative source of "revanchist" German aggression. Between 1961 and 1963, an echelon of 700 Medium and Intermediate Range Ballistic Missiles was built up within the USSR's main western border. Yet at the time of the 1962 Cuba crisis, the USSR had but 75 Intercontinental Ballistic Missiles (ICBMs) deployed in silos whereas the USA already had three times that number.[7]

Estimates on each side of adversary strength in this and other fields was much assisted from 1961–2 by the introduction, first by the USA and then the USSR, of reconnaissance by unmanned satellites in low Space orbits. May one add, too, that the American–Soviet disparity *vis-à-vis* intercontinental orders of battle looked at least as great when account was taken of strategic missiles installed in submarines.

The same applied the more if allowance was also made for the reaction times of the respective missilries. Thus, as its name implies, the solid-fuelled Minuteman ICBM the United States Air Force began emplacing rapidly from the Spring of 1962 could be very readily launched.

Mixed Messaging

On the Ann Arbor campus of the University of Michigan, US Secretary of Defense, Robert McNamara spoke thus in June 1962: "If, in spite of all our efforts, nuclear war should occur, our best hope lies in conducting a centrally controlled campaign against all of the enemy's vital nuclear capabilities".[8] He was alluding, above all, to those Soviet ICBM silos.

His averration was made primarily to refute Gaullist suggestions that the American guarantee to Europe had weakened because strategic nuclear equivalence supposedly prevailed already as between the two superpowers. Inevitably, however, the American enunciation thus of a Counterforce strategy up-and-running was also picked up in Moscow. The Soviet attempt that autumn to have nuclear-armed missiles emplaced in Cuba should therefore be seen as the Kremlin's urgent endeavour to offset a critical margin awhile of nuclear inferiority.

The Ann Arbor statement on pre-emption was intended to induce moderation all round, French and British military nuclear deployments certainly included. But it could have led instead to a nuclear exchange through miscalculation by a key player or players. Many if not close to all of us working then in this field, in or near government, failed to appreciate this adequately.

Between 1968 and 1982, the USSR conducted test interceptions with a liquid-fuelled Anti-Satellite (ASAT) homing rocket able to enter pursuit orbits at altitudes of up to 1500 kilometres. But a strategic Anti-Ballistic Missile deployment was conducted only around Moscow. Such remains the case. Intercontinental strategic parity was actually achieved, in the late 1960s, by the USSR deploying a sufficiency of Intercontinental Ballistic Missiles (ICBMs). Hence the Soviet–American Anti-Ballistic Missile treaty of 1972. It betokened unwonted acceptance of stability in a critical sphere.

Uncertain Outlook

As regards the Russian future *vis-à-vis* NEO defence, a lot may depend on how the folk traditions of the country are aroused. The Orthodox Church, perennially inclined towards conformity within the context of national separatism, has lately been waxing strong again with more than two-thirds of all Russians affirming at least nominal support. Prejudice against the heterodox – notably of late, gays and Moslems – has been an indicative downside of this.

As noted in Chapter 3, Alexander Solzhenitzyn urged his countrymen to ease up on their preoccupation with the outside world and focus on the "Russian North-East". Its "boundless expanses, senselessly left stagnant and icily barren for four centuries" need the sacrifice, zeal and love of "a free people with a free understanding of our national mission" to "awaken them, heal them, beautify them with feats of engineering".[9]

Alas, the portion of Siberia which qualifies reasonably enough as the "Russian North-East" is actually the worst for soil acidity, much worse than even the *taiga* plain. Seen from the air, it is largely comprised of some of the rawest mountains anywhere. The muddled romanticism and obsolescent "conquest of Nature" undertones that informed the

Solzhenitzyn disquisition quoted above were the *alter ego* of his savage critique of how the West was abusing democracy. The "continual seepage of liquid manure, the self-indulgent and squalid popular mass culture" he saw as threatening to undermine deeper values everywhere, Russia included.

But a popular reaction with a rather different thrust was soon to be observed within the Russian Far East itself. In the first 15 post-Soviet years, the defined region's population fell through emigration from 8.5 to 7.0 million. Lately Chinese immigration has been offsetting somewhat this trend, nativist resentment notwithstanding. But in 2016 Moscow began proffering modest Russian land gifts to prospective Russian settlers.

Also to reckon with is the regional impact of global climate change, given that Russia's weather has so long been characterized by the wintertime Siberian anticyclone sending drives of starry cold air across Europe. During the very moderate global warming of the early medieval period, however, there was a rather greater tendency for the winter atmosphere periodically to reorganize with an anticyclonic centre forming more often in high Scandinavian latitudes instead.[10] There are the signs this might be happening again; and, in any case, greenhouse warming could have negative consequences for ground drainage across much of the *taiga* zone. Apprehension of this could make Moscow the more willing to collaborate *vis-à-vis* stability in all its planetary aspects.

Meanwhile its cult of the frozen North is likely to stay in recession. In that regard at least, the contemporary Russian outlook may be closer to Peter the Great in his generation than to either Joseph Stalin or Alexander Solzhenitsyn. Building positively on this to work towards the creation of a multinational anti-NEO screen will likely wait upon the final departure of the Vladimir Putin generation of former apparatchiks and a consequent easing of tensions, in the Middle East and elsewhere.

(2) The Chinese Heritage

For a planetary survival strategy at all akin to the one herein propounded to stand a chance of success, it needs be especially cognizant of China's perspectives long term. Several times these last two centuries, the country has seemed liable to fall apart. Yet now she is poised, however uneasily, to become the world's largest economy, this well before 2100. Meantime, her rich scientific heritage is being invoked once again. Moreover, she has long known sectarian religious mixing of a kind liable to encourage a broad philosophical understanding appropriate to the Space Age. Thus "hundreds of millions of Chinese may be counted by the enumerators as Confucians but they might be, and sometimes are, counted with just as much reason as Buddhists or as Taoists".[11] This was penned in 1948 by a British agnostic/atheist imbued with a wholesome respect for enlightened religious persuasions. It may have relevance still.

Confucius, *c.* 551 to 479 BC, in effect became, through socratic discourse with his pupils, the world's first professor of political science. His understanding was nurtured by regional travel and debate. His analects were intended to prepare young men for government by inculcating within them moderation, courtesy and fair play. Since Mao Tse-tung's demise in 1976, the Confucian legacy has been drawn on, extensively and positively, by the authorities in Beijing. Taoism emerged in the third century BC as an indigenous development of Buddhism. Its emphasis has always been on simple living and plain talking.

Skilfully excavated and presented, the Neolithic village of Banpo in West China proclaims the continuity across the millennia of "we Chinese".[12] This heritage is similarly betokened by the imperial sky science annals, mainly meteorological but also astronomical. Nowhere else has Halley's comet been registered so regularly for so long, the earliest surviving record being from 613 BC. Moreover, by 1100 BC, a lunar-solar calendar is known to have been in use. From 104 BC, imperial calendars, lunar and solar, were regularly promulgated. Extant, too, in the Han was the concept of an infinite universe, with the stars floating in empty Space.[13] The constant though irregular efflux from the Sun of the sub-nuclear particles we know as the "solar wind" was picked up in AD 365, mainly through how cometary tails slew away from the Sun.[14] A gravitational perturbation during the AD 837 Halley fly-past is discernible from the surviving record.[15] Systematic sunspot readings dating from the Han have been used in a quest for millennial cycles in solar activity. One of 210 years is tentatively perceived. Witness a recent Nanjing study.[16]

Interchange of Buddhist scholars between India and China was well established early on. Several Indian astronomers and mathematicians were appointed to high positions in the Astronomical Bureau during the rather resplendent Tang dynasty (AD 618–907). One became President of the Board of Astronomy and hence masterminded a famous compendium. Links with Islamic central Asia were likewise to be important.

The well maintained meteorological records of imperial China have much facilitated our current interpretation of historical climate change, especially as regards the incidence of rainfall extremes. In 1926, the ratio of recorded droughts to floods was worked out century by century[17]; and the profile obtained was to be quoted in 1959 by Joseph Needham in his magisterial work on historical Chinese science.[18] Over the time span considered (AD 200 to 1400), only the fifth and fourteenth centuries seem in line with concurrent European experience. No less significant as expressive of Chinese scientific enquiry, however, was the limited concurrence duly recorded in this regard, century by century, between the two great river basins, the Yangtse and the Yellow River.[19]

Medieval Boom

But there were to be more striking examples of thorough Chinese scientific prowess. During the eleventh century, she embarked on an agricultural revolution which dramatically improved the adaptiveness of the farming sector to the conditions induced by early medieval global warming. The gist was the widespread distribution via Fukien (following an imperial edict of 1011) of rice that had originated in Champa, a temple state in the lower Mekong basin. Introduced initially because droughtiness had caused crop shortfalls on the Yangtse/Hwang-ho plain, Champa rice possessed, in one variety or another, two most positive attributes. The one was a pronounced ability to withstand water shortage; the other was early ripening. Though rather low in crop yield per unit area compared with what varieties long established on the alluvial plains optimally achieved, the Champa rices proffered an extra dimension to the national grain basket.[20]

The fact that this adaptation was demand-led related to the eleventh century being one in which the Chinese economy as a whole went through a market-driven lift-off. For a trenchant depiction, one is indebted to William McNeill. Though the imperial government still kept "the capitalist spirit firmly under control, the rise of a massive market economy from the 11th century may have sufficed to change the world balance between command and market behaviour in a critically significant way. China readily became quite the richest, most skilled and most populous country on Earth".[21] But one rider to this judgement could be that, being a soft grain, rice is ground into flour very little. So there may have been but limited watermill experience at village level.[22]

Kublai Khan

The Mongol struggle to subdue medieval China was long drawn out, initial successes notwithstanding. In its aftermath, Kublai Khan (1216–94), the able grandson of Genghis, evidently heeded a piece of simple advice from Ch'u-ts'ai, a Chinese adviser to his regime. This was that "The Empire was won on horseback but it won't be governed on horseback". Though asserting a notional suzerainty throughout the new found Mongol domain, Kublai made himself more particularly the first Emperor (1260–94) of the Yuan dynasty in China cum Mongolia.

His fostering of Chinese art and literature evoked a response. A fine tradition of mature painting was upheld. Vernacular drama and fiction enjoyed a pristine boom. Meanwhile Kublai renovated the pre-existing imperial infrastructure (roads, canals, granaries . . .). He transformed Khanbalik (the name he gave Beijing) into a truly imperial capital. He fostered foreign trade and diplomacy. In short, he did turn away diametrically from the "man on horseback" mode.[23]

Incipient Globalization

In some elementary though fundamental respects, China now seemed well-placed to lead the way to globalization. She also invented gunpowder and the magnetic compass. So did she the stern rudder and watertight compartments. She created woodblock printing though it was to be the Europeans who progressed to a general use of movable type. China had made paper since *c.* AD 100. Other Chinese inventions of the early Middle Ages included paper money, coke smelting, water-driven spinning machines, chaindrive and – in 1088 – an astronomical clock.

Above all, China led the world into the age of merchant adventurism. With three ships, Columbus left for the Indies, as he thought, in 1492. Yet in 1405, Admiral Zheng had led out of the Yangtse a fleet of 300 ships with a complement of 28,000. He thus began a saga whereby a succession of expeditions bespoke a modulated dominance within the Indian Ocean and, indeed, beyond.

As to how far beyond, a challenging hypothesis has lately been advanced by Gavin Menzies, an ex-Royal Navy submariner with a strong Oriental background. He has proposed that, between 1421 and 1423, four Chinese task forces traversed the High Seas, including the Arctic and the fringes of Antarctica. Among the manifold skills applied was the use of lunar eclipses in longitude determination.[24]

Withdrawal

Yet courageous navigators were soon to find, on their return from wherever, the imperial authorities turned off the whole idea. In fact, a voyage to Arabia in 1435 was to conclude the forays. Just how categoric this rejection became is shown by "the destruction of many of the records of the fleet and its ships' designs".[25]

The reasons still await full elucidation. But resentment at Zheng's virtuosity was probably compounded by his being a Moslem eunuch. Also a serious fire in the Forbidden City in 1421 (apparently due to ball lightning) will likely have been read as divine admonition. Meanwhile unrest induced by famine was becoming a problem yet again in North China. Added to all of which, some of the task forces had incurred heavy losses.

But behind the *volte-face*, one can also discern mandarin anxiety about cultural pollution. Perceptions were changing too fast within China and, indeed, elsewhere. Maybe the Chinese were wise in their generation and the West Europeans foolish. Thanks to the latter, a huge change of outlook suddenly ran across the world and continued strong. And then in a singularly fractious way, maritime Western Europe took all too full advantage of China's conceding pole position. Voyaging led readily to colonization, with Iberia in the van.

Any attempt at a more rounded explanation would extend well beyond the remit of this study or the competence of its author. But may one just say this much. Since 1985 or thereabouts, Confucius has been typecast as a primary source of the "Asian values" that had till then well sustained economic progress. However, such half-received wisdom stands sharply at variance with the judgement of Max Weber (1864–1920). This pioneering German political sociologist blamed China's material backwardness on the rigidities engendered by the integration of the Confucian literati into the imperial court, the past two thousand years and more.[26]

To Weber, the Confucianism he knew of was, like Roman Catholicism as he saw it, inimical to enterprise and liberty. Recently, however, a Chinese doctoral student at Columbia concluded that, in the developing world today, Catholicism can often be quite a force for progressive transformation. Likewise, his researches on twentieth-century China had indicated how often people informed by Confucianism had spearheaded earlier movements for democratic change.[27]

The Tao of Physics?

However, the debate thus joined has since extended well beyond Weber or, indeed, Confucius. In 1974, a study entitled *The Tao of Physics* (i.e. "the path of physics") came out, its author being Fritjof Capra of the University of Vienna and the Lawrence Berkeley laboratory. As a publishing venture, it was brilliantly successful. But was this prowess matched by the validity of its central thesis: that there was currently a movement in train in Western natural science (physics in particular) towards the "Eastern mysticism" of Buddhism and Hinduism?

Early on, Capra was criticized on several counts by Ian Barbour, Professor of Science, Technology and Society, at Carleton College, Minnesota. Barbour found Fritjof Capra too readily equated the unity which Asian tradition (in its mystical aspects and otherwise) has accorded the cosmos with what modern international physics in general knows of. The former is essentially undifferentiated whereas the latter is "highly differentiated and structured, subject to strict constraints, symmetry principles and conservation laws".[28]

Granted, Capra well identified many of the themes that so broad a cross-cultural comparison should be concerned with. For instance, he reminded us that "the Eastern image of the divine" was of a "principle that controls everything from within",[29] not from above. This concept is important in its own right. And here it was part and parcel of a well-found emphasis by Capra on the multifarious notions of consciousness.

Mysticism

Where one has to take issue, however, is over Capra's disposition repeatedly to draw the sharpest of distinctions between Western "fragmented" logic and an "Eastern mysticism" in which "for instance, the traditional concepts of Space and Time, of isolated objects and of cause and effect lose their meaning."[30] Episodic commendations of quantum mechanics and relativity theory do not validate more basic contradictions.

May one further submit, by way of reply, that a mystical sense has never been absent from Western culture, its scientific wing most certainly included.[31] Even during the late medieval times of doubt and disintegration, it waxed strong. One should particularly note, too, that Bertrand Russell, still regarded by many as the West's greatest atheistic logician, viewed the phenomenon empathetically, almost possessively. Mysticism, "embodied in theories of light", he saw in 1948 as originating in the West, apparently with Parmenides.[32] Indeed, as he freely conceded, "I yet believe that, by sufficient restraint, there is an element of wisdom to be learned from the mystical way of feeling which does not seem to be attainable in any other manner."[33]

Beijing Conference

In the year 2000, it was my pleasure to edit the proceedings of a conference held in Beijing that May under the auspices of the Oxford Research Group. The conference was attended by *c.* 30 ranking officials and academics (in post or retired) from China, Britain, France and Germany. Neither during the conference proceedings nor during the very interactive editorial process which followed did one discern even a hint of the basic cultural differences Capra would have envisioned. The Oriental and Occidental perspectives simply melded during this discourse which was concerned with sky science as applied to anti-missile defence.[34] We all freely moved from localized exercises in deterrence logic to broader general inferences and *vice versa.*

Yin and Yang

One could fairly say that this was because the two sides were heirs to the same tradition of metaphysical reductionism evident as from the earliest natural science, this emanating in each direction considerably from India. As much was apparent not least in the records of Chinese imperial astronomy cited above. A respect in which Capra could further be challenged was in his suggesting that this reductionism found singular expression in the Taoist notion of *yin* and *yang*. The *yin* principle (dark, maternal and receptive) is classically associated with the Earth. The *yang* principle (male and creative) is said to belong to Heaven.

Applied in traditional Chinese medicine, the inside of the human body was taken to be the *yang* and the outside the *yin*. A healthy balance was said to be kept by a continuous flow of vital energy, *ch'i* along a meridional system punctuated by acupuncture points. But such punctuation would implant in us an anatomy redolent of a trilobite fossil rather than what we all know well.

Yet however weak its clinical claims, the *yin yang* model can well be adapted to dialectical relations, ranging from the sex war within marital partnerships all the way up to Superpower showdowns, the latter scenario being one Revisionist Marxists have addressed to effect.[35] Modulated action–reaction is the warp and woof of social groups, parliamentary democracies, multinational alliances and, indeed, warlike crises. See Prefatory Statement. However, unbounded action–reaction could soon tear our fragile world asunder. So before attempting to explore further in context our evolving relations with China, we do well to recall who it was who advised that, if an enemy smites us, we might the better "turn the other cheek". He was not some oriental sage. He was Jesus of Nazareth. Subject no doubt to continued government monitoring and modulation, non-dogmatic Christianity may be a positive influence in China in the decades ahead, this in part through much resonance with *yin yang*.

Be that as it may, although the Chinese classically kept voluminous records of the skies above, they never built up a deductive interpretation of Creation after the manner of Aristotle, Ptolemy or Copernicus. Apparently, Confucius wrote but a single essay along such lines; and this, one gathers, was soon lost. The classical Chinese scholars rarely spoke "in terms of a God beyond the world, a maker of it".[36]

(3) China Ascendant

In the late eighteenth century, China was at what has yet proved to be her greatest territorial extent. Psychologically, however, she was introverted. In 1793, Britain sent her Ambassador MacCartney to negotiate a trade deal. The Emperor advised him that China had no need of further such progress. MacCartney was to liken his hosts to "a crazy old man-of-war . . . which . . . can never be rebuilt on the same keel".

Each brought grist to his mill. The Eurocentric world order by then emergent would be based on the principle of survival through competitive expansion. For China, in particular, this was going to mean over two centuries of foreign incursions with millions of derivative deaths along with orgies of dogma, deceit and general mayhem. As the nineteenth century wore on, China's fragile unity was especially compromised by the establishment about her coasts, and to quite an extent up her rivers, of no fewer than 80 treaty ports – locales in which privileges of trade and governance were extended to a foreign presence, this usually just the one European country. Correspondingly, Western values were permeating China more generally : Christianity, capitalism, social democracy, mechanization, medicine, modern science, consumerism. . . . Too well mixed a blessing.

Quite the bloodiest episode in China's violent nineteenth century was the Taiping ("great peace") rebellion, 1850–64. The credo of the rebel movement derived considerably from Protestant Christianity. The uprising began in South China but spread rapidly into and down the Lower Yangtse basin, making Nanking its capital. It was finally to be crushed with Western military assistance. The overall death toll is regularly put, however incredibly, at 20 million.

The Chinese Republic

In 1911, a few years after the legendary Boxer Rising, the last imperial dynasty gave way to army and business pressures in favour of a republic. A Western-style political radical, Sun Yat-Sen (1866–1925) was provisionally made its president. Come the First World War, 140,000 Chinese labourers lent logistic support to the Allied side in France, an ill-managed colonialist programme that gained for the infant republic little regard.[37] Then, during the twenties, Chinese radical politics became savagely split between Marxists and anti-Marxists, respectively led by autocrats then indomitably in their prime: Mao Tse-tung (1893–1976) and Chiang Kai-shek (1887–1975).

With the irrevocable Communist triumph on the Chinese mainland in 1949, most political progressives of my generation anticipated the emergence there of a rural socialism more evolutionary and pragmatic than what might have been forged behind urban barricades.

Yet *vis-à-vis* its two keynote programmes, the Great Leap Forward from 1958 and the Cultural Revolution (1966–74), the Maoist impact on ecology, economic development and, indeed, self-expression, was horrendous. Visiting Beijing in the late eighties, one found it was the Cultural Revolution that mostly came up in conversation. Those who had lived through it all told of rabid xenophobia, shattered universities, closed research centres, desecrated churches, crazed students, thousands of casualties in street violence, authors exiled to the remote countryside, high mortality rates in the corrective labour camps and gleeful destruction of cultural relics.

As regards the Great Leap Forward from 1958, its salient objectives were bizarre: create jumbo agricultural communes; make backyard "steel" producing plants; melt down kitchen utensils; kill all sparrows; eliminate flies. . . . Even Marxist sources accept that 20 million more humans died during the several years in question[38]; other reckonings go beyond 30 million. In 1960, Moscow was to end peremptorily a very diverse programme of economic and technical aid, hundreds of specific agreements being cancelled. Then some years later, she began what proved to be a sustained and very considerable military build-up along her long border with China.[39]

In the West, there was something of a disposition to interpret this reaction as but a continuation deep into the Khrushchev era of the

Stalinist precept of waxing hostile towards any foreign Communist movement or regime not well placed for overriding control from the Kremlin. There was certainly in the West a perception of the Soviets being anxious to preclude demographic spill-overs from China to Siberia, such as have actually happened on a significant scale these last few years. What the Soviets may have appreciated more readily than the West, however, is how contrary this Marxist deviation was to any sane definition of "scientific socialism" or, indeed, scientific anything. The Great Leap Forward especially afflicted North-West China.

More positively to be remarked, however, is how adeptly the Chinese proceeded with certain aspects of technological modernization, internal tensions and Soviet constraints notwithstanding. Take nuclear deterrence. They exploded a fission bomb (based on uranium enrichment) in 1964; and did likewise with a thermonuclear warhead in 1967. For some years, they evidently had big difficulties with building submarines fit to carry strategic nuclear missiles but since seem to have overcome these, sufficiently at least.

Post-Mao

Whatever view one may have taken of the implications of all this for the military balance of power, the nuclear technology advances were amazing in themselves, given the ambient circumstances. Indeed, they heralded a science and technology drive which was to strengthen remarkably in the aftermath of the death of Chairman Mao in 1976. China soon seemed to be into everything. Her consumer goods became famous the world over, not least for generally consistent quality control. Her economic growth percentagewise may well have been exaggerated somewhat over time. But there can still be little disputing its having been maintained for some decades at rates unprecedented for an ultra-large polity. This last year or so, the recorded annual increase has dipped percentagewise. But that drop, which may be partly due to diminished innovation, still leaves China with a growth rate other emergent economies might envy.

Nor has economic evolution been simply a function of product development. Since 1978, this has gone hand-in-hand with structural change towards what the Communist Party was to declare in 1992 was its goal of a "socialist market economy". In 1990, stock markets had been launched. In 1997, China was to begin the privatization or large-scale consolidation of many enterprises. In 2001, she joined the World Trade Organisation. The "Great Leap Forward" rural communes no longer exist, save for one preserved as a historical curio – a terminal slot in History for the deification of Chairman Mao.

Since 1998, there have been over a million village elections in China, not a few of them likely being authentic enough. Overall, however, reform and innovation in the Chinese countryside has been altogether too patchy, especially as seen in relation to disturbingly rapid urban growth.

A Crossroads?

Granted, Xi Jinping is launched upon an anti-corruption campaign which could prove stronger *a priori* than any previous one in the Republic of China. But the streamlined centralization President Xi has deemed needful to make her political economy work licitly and well is at variance with the decentralization that is the *sine qua non* of free market competition, not to mention grass-roots democracy.

The contradiction inherent in this situation will oblige Beijing to react decisively in one direction or the other, either back into even more uncompromising authoritarianism or else firmly forward to liberal democracy. Should she choose the former path, initially the easier, it will be hard to envisage her upholding the while a planetary strategy formulated to cope globally with major natural hazards. Yet without her involvement, any such claim to common purpose will lose a lot of its import. The West needs do whatever it can to stop this going by default. The Trump administration must be brought to understand this.

Military Transformation

Among the bureaucratic structures at least partially resistant to liberal reform is now the Chinese army. It is certainly one President Xi has felt obliged to confront very directly. By the end of 2015, several dozen generals were being held on graft convictions or charges. Illicit practices, common of late, have included bribes to secure promotion; and the redirection upwards of resources officially earmarked for the provision of rank-and-file housing. But lately Xi Jinping has been filling some of the top military posts when vacant with officers considered very much his *protégés*.

In September 2015, a huge military parade was held in Beijing, the first for six years. In the course of it, the President announced that reductions in military personal would be continued, this to the tune of a further 300,000. Yet neither he nor anybody else gave any indication when or whether the double digit annual increases percentagewise in the defence budget were to be reined in.

What does appear to be in the offing, however, is a reduction from seven to five in the number of military regions China is divided into; and the new regions will share resources and interchange personnel more freely than before. Moreover, a renewed emphasis, within the armed forces as a whole, on their ideological commitment, should further help President Xi curb irregularities in that sector.[40] So should his personal military connections. In addition, the Central Military Commission, chaired by the President, now exercises direct control over each of the customary service commands – army, navy and air. So does it over a new command for Space science and cyberwarfare as well as one for ballistic and cruise missiles.

Sky Science

Come the recent turn of century, sky science in general (and Space science in particular) was arousing much interest in China, not least among aspiring patriotic youth. November 2000 saw China's first White Paper on Space. One point this made was that gunpowder, "the embryo of modern Space rockets", was a Chinese invention. In October 2006, a second White Paper stressed that "China will focus on certain areas while ignoring less important ones". Themes to be prioritized were satellites, launch vehicles, telemetry, tracking and command facilities, manned Spaceflight including relatively deep Space exploration.[41] Around this time, China let it be known that she specifically aspired to be a lead player in a manned landing on Mars in the 2040s. Little has been heard of this recently but the aspiration is likely to remain extant.

China, like Russia, has not sought to preserve as tightly as the West does the peacetime distinction between civilian and military management of Space assets. This needs be borne in mind when citing numbers of military satellites in service. In early 2016, the three leading countries on the military side were China with 77 Space satellites; Russia with 80; and the USA with 122.[42]

A current cause for concern has to be that North Korea's progress towards a nuclear-armed and medium-range ballistic missile is making Washington feel the more obliged to offer surface-based Anti-Ballistic Missile coverage to South Korea and Japan. That in its turn could induce Beijing to accelerate sharply its development of strategic nuclear deterrence.[43] Pyongyang evidently continues to find in all this a congenial alternative to internal reform, economic and political.

Balanced Development

Chinese Space development in general will continue at a fair pace partly because Beijing remains rather more persuaded in principle than is opinion in the West these days of the "spin-off" benefits for economic development at large. A further link between grand strategy and political economy is the interest being shown in Beijing in the possibility of a new international reserve currency based on China's renmimbi currency and the American dollar. Beijing has overtly expressed the opinion that some such structure is needed.

At the back of everything is the question of the measure of autonomy Beijing may be prepared to concede long-term to Sinkiang, Tibet, Hong Kong and Taiwan respectively. Obversely, what autonomy is sought will depend the while on the progress Xi Jinping makes with the quality of government in China Proper.

Inspiration from the deep historical past will be on his side. Take the terracotta army first uncovered in 1974, since when it has been the subject of breathtaking revelation still very much on-going. It was largely the product of harsh forced labour, imposed those two millennia ago to try and alleviate the first Han Emperor's fear of the afterlife. But it remains peerless as regards vision, artistry, management and craftsmanship, the harshness notwithstanding.

In June 2016, President Xi announced that his government's policy was that China should become a leading scientific innovator by 2030 and a "dominant" science country by 2049. In terms of scientific papers published in English, she is already second only to the USA. Granted, too much of her present output is said to be either insubstantial or downright fraudulent. However, quality control has been improving.

She is also moving up-front with certain specialist installations and equipment. Thus she is developing further (for neutrino and dark matter research) what will become the world's biggest subterranean laboratory. Then again, from September 2016, Tianyan has been operational. With a diameter of 300 metres, it will be the world's biggest single-aperture radio telescope. The next largest will be one which stands at 200 metres.[44]

May one draw attention to a recent IISS study that discusses the Chinese cyber revolution with historical finesse: "China has spent much of the last 150 years attempting to catch up with scientific and technical advances of the developed world while maintaining a sense of national self-esteem commensurate with a state that, for most of recorded history, has regarded itself as culturally, economically and every way superior to other nations. From a technological perspective, China was for many centuries fully justified in this perception".[45] History teaching the world over needs take cognizance.

Still, the last word for now on China's drive into Space and all else may best come from the late Lee Kuan Yew, the brilliantly dynamic Prime Minister of Singapore (1959–1990) and a peerless blend of Straits Chinese enterprise and Cambridge legal scholarship. He said in 2011–12 that China's "reawakened destiny is an overpowering force". It was her "intention to become the greatest power in the world – and to be accepted as China, not as an honorary member of the West". India will not match her because she is herself "thirty-two separate nations that happen to be arrayed along the British rail line". Meanwhile "Westerners have abandoned an ethical basis for society, believing that all problems are solvable by a good government. . . . In the East, we will start with self-reliance". A sounder distinction than Capra drew?

8

Terrestrial Belligerence

(1) Near Space

My first contribution to what was then the nascent realm of strategic studies – namely, the 1960 Trench Gascoigne essay – dismissed any idea that Space could be set alongside Land, Sea and Air, thereby to constitute a fourth domain in war. All but sixty years later, as my generation marches steadfastly into the twilight, I go on the opposite tack. I submit that Near Space is just as distinct a domain for military analysis as are the other three.

Taking this latterday interpretation as the more appropriate to changing circumstances, one must concede that the mismatch in question is not inconsiderably due to my failure to appreciate originally how much the dynamics of war would change. However, it also owes much to a different assortment of criteria being adopted when according Near Space a special character.

In 1960, warships deployed from shore bases to contest the control of the sea. Warplanes flew from aerodromes most essentially to sweep the skies above. Soldiers went to war basically to secure territory. In the modern Near Space domain, suchlike challenges and responses still feature. But they may soon be widely seen as overridden by how Near Earth Objects (NEO) arrive randomly, this from deep in Space.

Arms Control

By 1960, world opinion was discounting the General and Complete Disarmament (GCD) casually favoured post-1945. A clear majority of commentators had decided that, in a world as riven as ours again was by conflicts of attitude and interest, GCD could be but a phantasmagoric utopia.

With this downgrading came a broad though contingent endorsement of two further propositions. The one was that quests for adequate deterrence and arms control respectively are best seen as twin aspects of the same theme: "not contradictory but complementary", to quote a study launched within NATO in 1966 by Pierre Harmel, the Belgian Foreign Minister. The other proposition was that mutual reassurance, through military and political intimations, is the *alter ego* of deterrence. The most evident gain from this approach was to be benign reciprocity in regard to strategic thermonuclear weapons as between the USSR and the USA.[1]

Even so, unstable equilibrium is undeniably a disturbing aspect of the Near Space ambience as such. A Space-faring polity might switch from *détente* to confrontation in close to real time. That is to say, it might intercept satellites in orbit, from the ground or Near Space, well within the hour. Alternatively, it might start jamming signals from orbital satellites, probably by using other satellites, in a matter of seconds. Currently, 1300 operational satellites are in orbit worldwide; and hundreds more will be added within a decade or two, many with a mass in the very low kilogram range. International regulation will become much more difficult. But if it falls too short, orbital chauvinisms will replace it.

One must also anticipate continual incremental gains in the performance of satellites, the small but also the large. Around 1990, one was being told by some in Washington that the satellite-based Free Electron Laser (FEL) "holds out most promise" for Ballistic Missile Defence (BMD) post 2010. The genre had originally been developed in the middle seventies, mainly at Stanford.

An FEL is not a laser in the strict meaning of that term. Coherent pulses of radiation are produced, not via a chemical lasant but through the interaction between an electron stream and a field of magnets. Pulse wavelength will therefore be a function of electron energy and magnet architecture. So an FEL, unlike a true laser, can be tuned, maybe to minimize the absorption in the lower atmosphere that so often obviates laser transmission.

However, as of 1990, orbital FEL development for the Strategic Defense Initiative (see below) was still two orders of magnitude short of the power flux deemed requisite. Soon, it was no longer being worked on in connection with BMD. Nevertheless, one has lately heard that the option is being studied anew, this as before with a view to application in 25 years' time. How far this is with BMD in mind is unclear.

The density of the Earth's atmosphere decreases rapidly with height. At an altitude of 80 kilometres, it is less than one part in 100,000 of what it is at sea level. A modern Intercontinental Ballistic Missile (ICBM) is likely to lift to around that altitude a warload consisting of several decoys and perhaps five or ten Multiple Independently Targettable Re-entry Vehicles (MIRVs), each one incorporating a thermonuclear warhead. This warload would be for accurate dispatch on diverse trajectories at about the same time as the rocket fuel of the parent ICBM was burning out. But, in principle, an orbital FEL could intercept the ascending ICBM, this shortly before the controlled dispersion of its warheads and decoys occurred.

The Launch of SDI

Thinking along these or similar lines was prominent in the backdrop to President Reagan's launching in March 1983 of the Strategic Defense Initiative (SDI), the programme to make nuclear weapons "impotent and obsolete" by becoming able to intercept them before target impact. As an absolute aim, the principle was clearly preposterous. As a means of diminishing the intercontinental strategic nuclear threat, it was open to critical review on a number of counts; and by 1987 had lost favour considerably in Washington, not least with Vice President George H. W. Bush, senior – then a key arbiter of policy in this field. On the other hand, surface-based weapons for the engagement of shortish range missiles in flight can already be reasonably successful. The Israelis demonstrated as much in their most recent war with the Palestinians.

Time out of mind, we have been told generals are always preparing to fight yet again the last war. What we should also ponder is how often arms controllers strive to prevent its breaking out. A quest is urgently needed on all sides for a philosophy of Near Space management that will adequately address cumulative Space debris, communications congestion, and the dual civil-military domain. It needs take due account in particular of how inter-state hostility may find Near Space expression in agile jamming as well as or instead of explosive attrition.

A debate does seem at last to be burgeoning, in Washington at least. Patricia Lewis of Chatham House has lately discerned a rift between (a) a Pentagon largely persuaded that the only way to preserve a semblance of order in this sphere is for the American military to remain within it stronger across the board than any potential adversaries and (b) a State department which apprehends this could stymie positive international collaboration in Space.[2] In particular, of course, active anti-NEO cooperation could be at risk.

This last consideration could prove critical. There may be no room in Near Space for both national weaponry and planetary anti-NEO weaponry. One says "no room" for geodetic reasons but also for psychological ones.

Fiery Dragons

We do well to bear in mind throughout how deeply anxiety about the skies above is embedded in our general culture. The fact that "death rays" have been a staple of science fiction since H. G. Wells and others shows how mere allusion to them can evoke melancholic folk traditions: lightning, the "evil eye", the basilisk's stare, the firedrake. . . . When rays are trained downwards from the heavens, evocation can be at its strongest. Even the prosaic *Anglo-Saxon Chronicle* told how, among the "foreboding omens" that "wretchedly terrified" the people of Northumbria in AD 793 were "lightning storms and fiery dragons . . . flying in the sky".[3] Deep folk memories may caution us about there being something unacceptable, morally and aesthetically, about the competitive weaponization of Space. If the "heavens were made for wonder, not for war", might not the Space domain be rather the way through to international arms control?

Current Military Applications

By early this century, satellites in orbit were signalling small corrections to surface-to-surface missile trajectories which were already deadly accurate by historical standards. And also, of course, satellites are widely in use for wireless relay, although these transmissions are not usually too difficult to jam. They have also been employed for quite definitive Earth photography (recognizable kinds of vehicles on roads etc.) for over half a century. Also quite widely for electronic eavesdropping or jamming.

All the military satellites currently in service are unmanned; and one knows of no plans as yet to introduce manned variants. Civilian satellites, likewise unmanned, may be used by the respective military as back-up for navigation, wireless relay and meteorological observation. As regards the "met" provision, no other facility can begin to match a satellite's ability to scan automatically temperature profiles from oceanic or land surfaces to the tropopause and well above.

(2) Regular War

The term "regular war" is here coined afresh to mean military contingencies which are not primarily nuclear nor biological nor cyber nor insurgent. Could not regular wars develop in the aftermath of major tsunamis, supervolcanic eruptions or global pandemics? Could not they develop regardless if there proves to be a deep shift away from the planetary patriotism this study aspires to help place on a more assured basis?

Aerospace Power

The USA's Tomahawk cruise missile of medium range (*c.* 1500 miles) has the capacity to loiter while awaiting electronic guidance to a target. The Unmanned Combat Aerial Vehicles (UCAVs) coming into general service before long for reconnaissance, bombing and interception will have remotely controlled manoeuvrability in very elaborate modes.

The air will thereby become more evidently than ever the domain of war most radically effected by electronic advance. For received opinion seems now to have it that UCAVs are likely to be, well by 2050, the dominant vehicles in the leading air arms. Quite a few may be controlled from high up the command chain which can mean across great distances. Some Reaper UCAVs Britain's Royal Air Force have lately been using in the Middle East have been directed operationally from Nevada.[4]

Sea Control

In the sea domain, it will evidently be feasible to exploit the electronic revolution, most notably by aiming to reduce vessel size and to weaken submarine acoustic concealment. The most significant opportunity of all may be that presented as planes on aircraft carriers with long flight decks are replaced by UCAVs dispatched from launch pads on much smaller ships. Drones, crew-less surface vessels and submersibles will be able to deploy, not least against enemy submarines, more and better sensing instruments than maritime forces have in the past.[5] Obversely, though cargo liners will henceforward be almost fully automated, ultra-large designs are actively being contemplated.[6] These may sometimes have a crisis role.

Indeed, disaster relief is an aim which, in the decades ahead, is liable to figure contingently on a scale only sea lift can cope with. Several times this author has called for the development of a doctrine of Fighting Relief, a concept for protecting actively the delivery of relief aid ashore and its subsequent distribution. Amphibious craft could often play a special part in such contingencies as well.

Meanwhile, the open sea itself will require more routine protection. Thus there will be a growing need to police the exploitation of seabed oil deposits. Also to monitor rising acidity levels in the oceans. The Scripps Institution of Oceanography reported a fall in oceanic Ph readings worldwide from 8.11 to 8.07 between 1990 and 2013. This connoted a considerable rise in acidity levels. How much this matters is not yet well determined. But it could be quite a lot. More extensive quiet zones might also be needed to inhibit overfishing; and to shelter marine life in general from the noise of ships' propellers.[7] Yet acoustic jamming may also come to play a large part in submarine warfare in particular since simply keeping still and being quiet will not proffer concealment beneath the waves the thorough way they once might have.

A sense that pursuit of a variety of oceanic interests was somehow "domesticating" the traditionally "open sea" was reinforced in 2015 when the United States opened seven ocean observation stations to link up with three pre-existing Canadian ones in order to maintain at each locale a time series of many marine properties, this down to the sea floor. That same year, a valuable contribution in Britain's leading national journal on defence matters proposed that Japan's recent disaster experience (the Great East Asia Japan earthquake in March 2011; and typhoon Haiyan in November 2013) had shown the customary naval ethos still to be well adapted to civil crisis response, not least multinationally.[8]

The Uncertain Land Domain

Given that we humans are essentially land animals, the land is the domain that, in principle, most calls for war doctrine. And yet intricate geographical variations make the overall input of new technology on land war especially hard to assess.

The use of road bridges illustrates this. If firing just one missile can proffer a close to 100 per cent guarantee of denial until the bridge in question has been replaced or repaired, might this not mean that most military ground vehicles liable to be used in riverine landscapes should be amphibious? Additionally or alternatively, how much extra reliance must therefore be placed on helicopters?

Further questions stack up. What might be done in advance to lessen the impact of a precision surprise attack? Will minefields be needed after all? For might a field laid be collectively switched on or off? To what extent may exchanges of gunfire over open sights continue and how far will these be with bullets rather than shells? How wide and deep will the battle zones held by major formations tend to be? At what level of unit command (company, battalion, brigade . . . ?) should the integration of infantry, armour and artillery take place? Will this integration be *ad hoc* or on-going? What kinds of armoured vehicles will be needed and in what proportions? How far should battlefield information flow be processed locally?

Commonalities

Finally, one needs identify a mix of *a priori* questions which apply in good measure in respect of each of the operational domains. How much armed confrontation may occur? Might casualty rates often or sometimes be very high? In particular, will not the dividends from striking first be bigger than ever? So can states just rely on career servicemen plus volunteer reserves or must conscription sometimes figure? And where might battlefield nuclear weapons fit in, contingently at least? Saddam Hussein occupied Kuwait in 1990. Had he moved on forthwith towards the Saudi oilfields, it would have been very difficult to stop him without nuclear fission recourse.

Especially pertinent from here on out will be the actual or likely effect of much faster and more voluminous data flow plus, contingently, much higher casualty rates. How sharply might a contrast be felt with the traditional military emphasis on straightforward but categoric procedures? Will morale be more prone to snap? Will war crimes be more liable to occur? And what of the future for the officers versus other ranks binary distinction that seems invariably to be sharply preserved in armed services, more or less regardless of the wider political culture? In particular, should it be maintained within Cosmic Command?

Then again, how will attitudes towards making in youth the supreme sacrifice be affected when the alternative prospect for them may be a goodish life till well past a hundred? What kind of armed conflicts might arise in the aftermath of major natural disasters? What if they do in the aftermath of an NEO interception that has been unsuccessful?

Might crisis escalation sometimes be restrained by reverting to the custom of declaring war, perhaps war in a specific mode or with declared aims? How do we cope with conflations between insurgency war and apolitical violent crime? Should not biological bombs always be treated as weapons of mass destruction? What of chemical weapons? And should not a Cosmic Command assume some operational responsibility for terrestrial peace restitution – e.g. the collection of small arms? And can cyberwar be constrained satisfactorily?

Danger Inherent

Such a mix of uncertainties could induce states to arm themselves more heavily than they otherwise might have. Stockpiles of automated systems will accumulate over time in any case. This is bound to lead to misjudgements about what the right defence structure is for polities individually. All of which makes it the more imperative that we soon start to build a planetary security structure, one with two major settings – terrestrial and Near Space. And a stronger awareness of action–reaction will need to figure prominently in Earthly war avoidance or early termination.

(3) Insurgency and Proliferation

Insurgency and the proliferation of Weapons of Mass Destruction (WMD) can be respectively regarded as on the lower and higher margins of "regular war" as here defined. The risk is real that, 20 years hence and beyond, insurgency at least will be more prevalent across the world at large, even if the Middle East is more at peace by then. For such a trend might well be engendered by an inchoate form of Limited World War, one which (a) is not masterminded by a citadel state or a self-consumed charismatic and (b) melds into gang crime and "lone wolf" randomness. A root cause may well be a tendency for developments in war making to give more purchase to a few but less thereof to many, a military subset of the Olson dictum. See Chapter 3.

A secondary influence could be an uncomfortably positive correlation between economic progress and ecological spoliation. A tertiary effect might be an ever more insistent weakening of community relations. A quaternary one could be even less certainty throughout about general aims and purposes longish term. And might not these several tendencies readily conflate, resulting *inter alia* in dangerously deep divides between factions or, indeed, generations?

Related to some merging between organized crime and political violence could be the furtherance of a tendency already in train within insurgency – less reliance on inspirational guidance from singular supreme leaders – your Joseph Titos or Mao Tse-tungs. Strikes aimed specifically at such individuals by commandos or drone aircraft are contributing to this development. But an underlying reality is that insurgency is suffering from philosophical exhaustion even more than constitutional democracy is. Supremacists need philosophies, however perverse these may be.

To the erstwhile young progressives of my generation, myself most certainly included, Mao came across as pragmatic but caring, a likeable contrast with Stalin or Lenin, not to mention Chiang Kai-shek. Yet since his "Great Leap Forward" from 1958 to reform the Chinese countryside, he must instead be seen as an all-devouring maniac. See Chapter 7(3). Continental rainfall cycles tentatively traceable back to Han times[9] may have played a part in this disaster but, in any case, should have duly been allowed for.

Warranted condemnations will always assist the struggle against insurgency overall. However, that struggle remains liable to be compromised by its own ambiguities. Among the more obvious can be its often being waged not just by recognized democracies but also by regimes whose democratic credentials are weak to non-existent.

Privacy

One must say, too, that the amassing of personal data by the state and other institutions flies in the face of the principle that privacy is a cornerstone of individual liberty. Data collection and integration is constantly being diversified, refined and expanded: automated cameras and electronic eavesdropping; reconnaissance from the surface, air or Space; biological analyses; "virtual reality" creation . . . Moreover data flows are regularly at risk from cyber-hacking, not to mention old-time theft. Meanwhile the human operator is at risk of data overload, this liable to compromise his or her judgement and assurance. This especially applies where war is actually waging.

Biowarfare

A distraction within strategic studies currently is the propensity not a few of its acolytes have acquired for arguing the case for a complete abolition of nuclear weapons without checking out beforehand the prospects for ridding the world of biological bombs or, to be even more sanguine, any capacity for producing them. One does well to heed an admonition by the late Herbert York, first director of the Livermore (now the Lawrence Livermore) national laboratory and later the scientific adviser to President Eisenhower. To him, the biobomb was "the thing on the horizon. We in the United States in the name of examining what the possibilities are, *so we can defend against it*, are going to lead the world into a *much greater trap* than nuclear fission" (my italics).[10]

As to why the biowarfare problem has publicly been addressed so inadequately, it is likely that an apprehension is abroad to the effect that to talk up the subject could be to give its use a kind of *ersatz* legitimacy. Such an apprehension would mesh with a perception that the biowarfare community as a whole tends to be introverted, too preoccupied with the intrinsic aspects of its work and too assured as to wider connotations.[11]

Bioweapons could be of critical utility to guerrillas of otherwise limited means. At times, indeed, they could swing the operational advantage strongly in insurgency's favour; and that might similarly apply to terrorist rogue states. Several thousand biotechnology firms are spread around the world. There are no indications thus far that this particular proliferation owes much or anything to a desire on anybody's part to maintain an offensive biowarfare option. However, something fairly close to that could follow *ipso facto*.

Insurgent biowarfare development might only require modest premises (e.g. a domestic garage) which look commonplace from without. Much of the requisite background science could well come from blue skies research. The distinction between defensive and offensive preparations could be exceptionally hard to perceive. Likewise, the other side might well not be able to tell beforehand (except, just possibly, via espionage or defections) what sort of microbes might be used offensively. Indeed, the side under threat might be uncertain whether a sudden epidemic had been deliberately started.

All in all, grim dividends could accrue from a sudden first strike, be it aggressive or pre-emptive in motivation. One should add that, in so far as meaningful comparison can be made, the psychological impact of recourse to biowarfare might compound considerably the physiological. And the ensuing conflict could well be one in which initial advantage was everything. A planetary network of on-site inspection may, in due course, be looked for to preclude biowar development.

To which one should add that some authorities on biowarfare are or have been stressing how much harder crop plants are to protect against this threat than are ourselves or other animals.[12] What must be said, too, is that the risk of pandemics (human or otherwise; deliberately induced or otherwise) spreading is growing markedly on the shrinking planet where we find ourselves to be.[13]

Analogues as Precedents

What should also be acknowledged is how close a precedent is created by gas warfare, lethal or otherwise. Decades ago, a cat sprayed with a modicum of psychochemical gas literally became running scared of mice. Not too inhumane but very manipulative. Not the way we would expect any democracy to respond to hostility. Much the same applies to, for instance, the widespread use of Agent Orange as a jungle defoliant in Vietnam. Then again, as with the Viet Cong during that conflict, the international narcotics trade can be the source of another bio form of area attack. As a bold and brilliant Hong Kong-led study has recently shown, North Korea is currently the lead practitioner.[14] She seeks to gain foreign currency but also inside information and a corrosive influence.

Aspects which still stand in need of elucidation concern the operational relationship between Pyongyang officialdom and private traffickers. However, it is already evident that large quantities of North Korean illicit narcotics are being trafficked internationally by a gangland network active in both the Koreas plus North-East China. The South Korean and Chinese police have lately been collaborating closely on this account.

Soviet Deceit

Despite having ratified in 1975 the Biological Weapons Convention of 1972, the USSR had built up by 1989 a biowarfare network an order of magnitude more extensive than Western intelligence had suspected. Defectors were indicating that there was a Biopreparat "archipelago" of 15 major centres spread across the Soviet Union and employing 35,000 people.

In 1992, Moscow overtly acknowledged that its research into biological warfare had explicitly included an offensive dimension since 1946.[15] On that score, one could allow that, for the first twenty years or so, a modest programme along these diabolical lines could have served to offset the American and Western supremacy still prevailing then in nuclear fission and thermonuclear respects. However, that the Biopreparat was allowed to flourish subsequently has surely to be put down to Kremlin appeasement of a not so very strong military-industrial pressure group.

The Biowarfare Outlook

Our keeping quiet about biowarfare would not ensure its going away. Nor may those who disseminate microbes have too often to take the risk that dissemination will go full circle and spread back to infect themselves. Sometimes, indeed, the most effective mode could be to dispatch to the target area a colleague who has readily accepted infection – an undercover *kamikaze*.

Microbes may go through a million generations in the time humankind takes to live out one, a contrast which tends to make the former far more adaptive, in a state of nature at least. Furthermore, it will be quite infeasible, logistically and physiologically, to vaccinate whole populations against the scores of pathogens that might be 'weaponized'. Also to reckon with, in due course, could well be the synthetic creation or adaptation of microbial life to obtain more aggressive variants.[16] A particular option may be the generation within certain microbes of genes that will produce deadly toxins.[17]

Nuclear Proliferation

However, one is also aware of a burgeoning apprehension that "nuclear proliferation" (i.e. the acquisition of nuclear weapons by countries that do not currently possess them) will soon be on the increase again. It is hard to say what this is mainly down to. Might this perception essentially be a pendulum swing? We were all anxious on this score two generations ago. So perhaps the time has come round again.

Besides which, though, there has been some shift of informed opinion back towards the view that without some continuing reliance on nuclear power, world energy demands will be hard to meet in the coming decades, whatever the circumstances otherwise. May one again underline, too, the prospect that a stratospheric dust veil consequent upon a largish NEO impact or a big volcanic eruption or nuclear war could globally obscure the Sun for months or years. Nuclear power readily in place could not obviate the effect on Nature at large but might ease somewhat the direct impact on ourselves. See Chapter 3.

South Korea

Granted, the International Atomic Energy Authority (IAEA) in Vienna has evolved an elaborate regime intended to preclude military nuclear proliferation to further countries. However, experience with the Republic of Korea (i.e. South Korea) has shown how hard any provisions made may be to sustain indefinitely, even when there has been no great pressure to overturn them. The record sixty years after Hiroshima and Nagasaki was to be scrupulously reviewed in 2005 in the *Bulletin of the Atomic Scientists*, a literary legacy of World War Two that regularly merits attention.[18]

Its findings can be summated thus. The Republic of Korea had conducted chemical uranium enrichment from 1979–81; separated quantities of plutonium in 1982; experimented with uranium enrichment again in 2000; and manufactured depleted uranium munitions from 1982 to 1987. International obligations notwithstanding, the Seoul government had declared none of these activities to the IAEA.

Nevertheless, the Korean Atomic Energy Research Institute (KAERI), based in Seoul, had consistently complied closely with IAEA safeguards as regards power reactors and related facilities. Besides which, many of those KAERI researchers then pushing beyond prescribed limits South Korea's large programme of civil nuclear power were fervently expressing opposition to any military nuclear dimension.

But should not this be seen in the context of American forces throughout remaining in strength in the Republic of Korea? Jungmin Kang and his *Bulletin* colleagues averred that it was "commonly known" that, from 1968 to 1975, South Korea attempted to obtain a facility for processing plutonium from spent fuel and also to procure intermediate-range strategic missiles. Moreover, from 1971, an effort to develop a South Korean nuclear bomb enjoyed presidential-level backing within Seoul, a short while at any rate.

These pitches were motivated, at least in part, by President Richard Nixon's unilateral decision to withdraw one of the two American army divisions in-country. Secretary of State Henry Kissinger put heavy pressure on the ROK to terminate this quasi-unilateral strategic programme, pressure which also provided an opportunity for American reassurance. In early 1975, Seoul acceded to the Non-Proliferation Treaty (NPT) and later that year signed a "full-scope" safeguards agreement. But as of the autumn of 2016, a majority of South Koreans were polling in favour of their developing "the bomb". Action and Reaction was playing in.

North Korea

In 1994 North Korea was to respond to IAEA evidence of undeclared activity by expelling IAEA inspectors. Then in 2003 she withdrew from the Non Proliferation Treaty. Pyongyang's especially bitter authoritarianism was and remains rooted largely in how Imperial Japan's colonization had been succeeded by a Stalinism interactive very negatively from 1945 with the fascistic regime that had come to power awhile in post-war South Korea.

Still, the bitterness cannot but relate also to the United States Air Force's very comprehensive bombing campaign, Operation Strangle, waged the last two years of the Korean War (1951–3). The aim of depriving the Communist frontline of ammunition resupply was only partially successful. But the impact on civil society was horrendous. Now Pyongyang's development of strategic ballistic missiles with nuclear fission warheads is being sustained by (a) the regime finding this comes easier, initially at least, than internal liberalization and (b) the operational progress being made.

Thermonuclear Proliferation

As 2015 drew to a close, multinational impatience with the quest for peaceful application of nuclear fusion was once more waxing strong. Granted, significant gains in computer power and precision engineering were in train or in prospect. Work on the Large Hadron Collider at CERN had also been quite relevant and encouraging. Even so, there were still few, if any, informed commentators who foresaw post-ITER[19] inputs into national electricity grids this side of 2050. Any unilateral diversion to explosive purposes would be unlikely for some years afterwards.

Yet innovation was in the air. Some analysts were pressing for fuller investigation of design concepts that allowed of magnetic confinement via smaller reactors than the tokamak mode can allow of. Once again, the stellarator was especially mentioned. Other analysts were looking towards more involvement by private enterprise. If either or both these trends do progress, they will favour a wider proliferation of smaller reactors. That might have strategic implications within several decades.

The late Herman Kahn and his Hudson Institute became well known in the sixties for their trenchant views on international security. Not least remarked on was his envisaging of a Doomsday Machine: a large thermonuclear warhead or battery of such warheads designed to be detonated *in situ* – a planetary *kamikaze* affecting neutral states as readily as belligerents. Kahn reckoned one such fixture could be built somewhere or other within a few years and for a few billion dollars.

Knowledge of nuclear fusion will spread more widely, just as physical science as a whole is doing. More specifically, all the polities in the ITER consortium (i.e. China, the European Union, India, Japan, Russia, South Korea and the United States) will learn a lot from involvement, whether or not individually they already have military thermonuclear status. Then again, on the military side, several states have moved historically from nuclear fission to nuclear fusion status in well under a decade. And now the Republic of Korea, in particular, is bringing to ITER nuclear fusion her extensive understanding of nuclear fission. Yet having discussed civil nuclear fusion in various operational venues across the years (i.e. in Birmingham, Canberra,

Culham, Naka, Osaka and Princeton), one has to say that military diversion never entered the agendae.

But this does not mean that thermonuclear warheads need not figure in the anti-NEO debate. All else apart, they may have a critical part to play in the interception of large NEOs. A "hydrogen bomb" in the megaton range may be essential to the effective deflection of, say, a billion ton comet. See Chapter 10(2).

9

Near Earth Objects

Crucially to reckon with throughout are those material objects to skywards which are neither part of what distancing makes an apparently immutable firmament nor regular enough to be numbered among the planets – the *planetes* or wanderers and their moons. The allusion is to what we know as Near Earth Objects (NEOs): comets, asteroids and meteors – meaning, in particular, those which seem liable to collide with our native Earth, sooner or later. A critical distance as regards this hazard has been taken as 0.3 Astronomical Units from the Earth or 1.3 from the Sun. An Astronomical Unit is the mean distance between the Earth and the Sun's centre. It is very close to 93 million miles. Usually, an NEO is understood to be a body believed to have an orbital period, one of less than two centuries.[1]

As indicated in the preface, estimates of the destructive force of NEO impacts are expressed in terms of the nomenclature originally adopted to cope with nuclear explosions. What needs now to be emphasized is that the extent of the material damage from a single impact or blast does not scale up anything like as readily as the kinetic energy thereby released. As is all too well known, in World War Two, terrible damage was wreaked across and around much of the Eastern Hemisphere and adjacent seas by the expenditure, in millions of rounds of bullets, bombs, shells, mines . . . , of an estimated six megatons of high explosive (plus or minus the forty kilotons via the two nuclear bombs). Then come 1961, there was to be that single nuclear fusion device reckoned at 60 megatons of energy detonated at the Novaya Zemlya testing ground on the USSR's Arctic seaboard. On its own terms, it was a fearsome event. Yet it could never have caused anything remotely like ten times the destruction and suffering humanity saw fit to inflict upon itself and the world at large between 1937 and 1945. Indeed, it could never have brought about anything like as much mayhem as one megaton of ordnance diversified to that end.

It might have come closest to equivalence through medium-term disturbance of the global ecology. An NEO more than, say, 200 metres across could well cause an oceanic tsunami if it hit fairly shallow but open sea. Witness, too, the spate of scientific literature some thirty years ago on how an all out thermonuclear exchange as between the Superpowers could well involve the expenditure, in the form of thousands of warheads, of five or ten thousand megatons of nuclear energy, causing *inter alia* in the rainless stratosphere a fairly persistent dust pall worldwide. This would likely induce a dramatic collapse (maybe ten or more degrees Celsius) in the global mean temperature – a "nuclear winter" from which complete recovery could be but gradual, years rather than months. The debate thus engendered was led, in effect, by the cosmologist Carl Sagan and by Andrei Sakharov, each arguing against a less apprehensive Edward Teller. He and Sakharov had been dubbed, with some justification, as respectively the "fathers" of the American and Soviet hydrogen bombs.[2, 3]

A 1908 event which then passed virtually unnoticed by the world at large but has been much remarked on of late was the NEO disintegration, at an altitude of eight kilometres, above the Tunguska valley in Siberia. The stony NEO involved had originally been, it is now believed, just 30 to 50 metres across. But as it broke up, the blast flattened remarkably evenly some 2000 square kilometres of *taiga* woodland. One source this study is drawing on, in this and other connections, put the Tunguska energy release at 10 to 20 megatons.[4] Another had allowed of 20 to 40 megatons, this reckoned mainly on the basis of barograph blips across quite a wide area.[5] But a recent dynamic event speaks just of several megatons.[6]

Scaling Up

The first assessment here referred to focusses mainly on a compilation which translates the diameters of hypothetical NEOs incoming at representative speeds into the kinetic energies predictable on their respective impacts. Also recorded is a guestimate of the average time between visitations for an NEO with a diameter within a given bracket. The said compilation covers a spread of NEO sizes (asteroidal or cometary) from below 75 metres in diameter to 16 kilometres and over. May one now use as markers four segments of the table in question.

From 75 to 159 metres across, the kinetic energy imparted by impact is given as in the 10 to 100 megatons bracket; and the mean time between successive such impacts is put at several thousand years. Between 350 and 699 metres, the megaton range is given as 1000 to 10,000, the mean incidence being cited as 16,000 years. A diameter between three and seven kilometres is associated with the energies respectively imparted being between one and ten million megatons; and impact is deemed averagely to recur every several million years. With an NEO diameter of 16 kilometres or more, any kinetic impact is thought unlikely to be less than 100 million megatons, an event bound to come close to extinguishing life on Earth. However, a typical recurrence time is put at 100 million years or more.[7] Suffice to add how it seems that, four-and-a-half billion years ago collision with something the size of a small planet caused our Moon to break off from our Earth and go into orbit round it.[8]

Comets

Historically speaking, the comet has been quite the most remarked of the NEO genre. Its identity rests on its being comprised of loosely bound material, icy and rocky, which exhibits a glowing head and extended tail or tails once it enters the inner solar system. During the AD 1456 return of what we come later to know as Halley's comet, Pope Calixtus called for noonday prayers beseeching God to deliver us from "the Devil, the Turk, and the Comet". His Holiness thereby encapsulated an abiding belief throughout various religious cultures that the singular appearance and eccentric transits of comets showed them often to be signs of anger in heaven though whom with or what about might not always be clear.

Several per cent of the comets in the solar system reside within the Kuiper Belt, a near to circular region beyond Neptune's elliptical path and between 30 and 50 Astronomical Units (AU) from the Sun. Planetary gravitational influences and/or collisions acting on the internal edge of the belt sometimes cause individual bodies to leave that ambience, quite possibly to enter upon orbits bringing them closer to the Sun. Their periodicity may then be less than two centuries; and they usually orbit in planes close to that within which the planets move. Both these attributes make them less hard to intercept. However, their speed and poor definition could remain big problems.

The soft landing of the European Space Agency's Rosetta probe on Comet 67P in September 2016 demonstrated very accurate steering. This was the culmination of a ten-year remotely controlled mission, up to 500 million miles out.

As the giant outer planets formed up, much of the icy debris whence their materials came was ejected outwards. Many of the bodies thereby created assumed very long solar orbits thereby forming together the Oort cloud, usually described as the shell of a hollow sphere containing billions of comets (many of them quite tiny) mainly 1000 to 30,000 AUs from the Sun. Occasionally a solitary comet is deflected from the cloud by gravity and/or collision. Entered thus upon a new orbit, it may pose an incremental threat to the Earth. But its orbital period may have to be guestimated in thousands of years.

Comets move at up to 70 kilometres a second as against the 15 to 25 typically registered by an asteroid, the primary NEO category we consider below. And, as indicated above, other things being equal, twice the speed means four times the kinetic energy. As a comet diverted by collision moves into the inner solar system, warming will allow much gas to escape and form around the solid nucleus a gas cloud, the *coma*. The *coma* of a comet nearing the Sun may visibly be 100,000 km across; and the relatively compact nucleus, 10 to 15 km. The tail may be as much as one AU in length; and may be telescopically visible in slanting sunlight five or ten AU away. As of the present, only one per cent of the observable NEOs are designated comets.[9]

Occasionally a gravitational deflection may impel a comet to leave the solar system[10], accelerating markedly the while. All in all, the number of cometary appearances registered by us Earthlings, century by century, does not seem all that variable, these last two millennia.[11] But individual size can vary dramatically. Some comets appear to have a mass of several hundred thousand or even a billion tons. Much of this will be ice.

The Asteroids

Between Mars and Jupiter (two to four AU from the Sun), the overwhelming majority of the asteroids long remain in solar orbital residence. Although the largest can be very broad in the beam (several hundred kilometres across), they may sometimes be spoken of as "the minor planets". Perhaps a million of them are over a kilometre wide. Their relatively low aggregate mass yet large number relates to their proximity to mighty Jupiter. The genre was first identified rather over a century ago.

Asteroids emit no visible light and very consistently present a darkish surface. Individually, they may be loose aggregations of material or else just the one solid lump. About three-quarters are carbonaceous, their chemistry being somewhat organic. Most of the remainder are silicon-based and rocky. But several per cent are metalliferous – iron, nickel, platinum and so on. Debate continues as to how these ores might be exploited, eventually. Debate is getting livelier, too, about the bound water content of certain asteroids. Viewed as NEO targets, asteroids are decidedly heterogeneous.

An especially daunting example of NEO impact was asteroidal. The allusion is to the Chicxulub event we see as defining the "KT" divide, some 70 million years ago, between the Cretaceous and the Tertiary geological periods. Come 1980, the conclusion was that this event was effected by an asteroid ten kilometres across impacting on the Yucatan peninsula of Mexico.

It accelerated sharply the demise of the dinosaurs except in that, since 1975 or so, strong skeletal indications have been that birds are descended from certain of them. Insect species largely survived as did most of the large amphibious reptiles, sharks, rays and skates. Plankton mostly died out in seas made too acidic. Mammals were to multiply and diversify strongly. The kinetic energy of this Yucatan impact was initially put at 200,000 megatons.[12] A higher guestimate might be made today.

Meteors / Meteorites

A meteor can be defined as a small interplanetary body, though one said to be older sometimes than rocks on our Earth's surface. A meteor that enters our atmosphere and, before its glowing ablation is complete, hits the Earth's surface is spoken of as a meteorite. Ninety-three per cent of the meteorites inspected have proved to be stony while six per cent have been predominantly ferrous. Meteors/meteorites remain subjects for passive defence, not active interception.

Sometimes a comet will shed meteors around itself and/or end up as a meteor shower. In June 1975, the Earth and Moon passed through a strong meteor stream. As regards the airless Moon, this meant more boulders weighing a ton or more hitting the surface in the course of five days than had done the previous five years. Virtually all the meteors entering the Earth's atmosphere disintegrated forthwith as per normal.[13] For a contrary terrestrial instance, we may look to the Britain of 17 centuries ago.

The Ruin of Britain?

Received scholarly wisdom used to have it that the human population of Roman Britain peaked somewhere between two and six million but that, whatever the actual maximum, it will almost have halved by the year 650. Today the extent of any decline is under review, albeit with information which remains patchy and confusing.[14]

At times, indeed, it seems plain contradictory. Take Gildas, a sixth century Romano British monk who resided in West Britain. His overriding concern was with the divine wrath facing any society which failed properly to respond to the beneficence of God. His celebrated tract, *De Excidio Britanniae*, carried in its preface a roseate account of his native land as it once was or ought to have been. May this extract suffice:

> "Poised in the divine scales that (we are told) weigh the whole Earth, it stretches from the south-west towards the northern pole. It has a length of 800 miles. . . . It is fortified on all sides by a vast and more or less inaccessible ring of sea apart from the straits to the south. . . . It has the benefit of the estuauries of a number of streams and especially two splendid rivers, the Thames and the Severn. . . . It is ornamented with 28 cities and a number of castles and is well-equipped with fortifications – walls, castellated towers, gates and houses whose sturdily built roofs rear menacingly upward."[15]

Yet this rhapsodic celebration relates awkwardly to the author's avowed aim of delivering, not a moment too soon, a jeremiad against a "general loss of good, a keeping up of bad" within society.[16] It reads the more awkwardly as a precursor to the account he rendered of the advent of the English. He acknowledged how stressed the Romano-British confederation had been by continual barbarian sorties, overland or amphibious. Even so, he deemed it terribly wrong to have invited in the Saxons to help defend the south-east coast. He saw them as graceless interlopers, somehow responsible for there now having been a more generalized conflagration:

> "In just punishment for the crimes that had gone before, a fire heaped up and, nurtured by the hands of the impious easterners, spread from sea to sea. It devastated town and country round about and, once it was alight, did not die down until it had burned almost the whole surface of the island."[17]

By what we wish to see as our standards today, this account was manifestly a blatant exaggeration of whatever did happen. But this was an age of very disparate and slender communication long-range. You often felt obliged to overstate massively in order to stand a chance of being heard at all. We today tend to go the other way about. We may feel confident we have accurately received hard facts about some far-flung incident. But we then treat it as more representative of reality than it is or ever could have been. We may do so especially if the incident seems adverse, a disposition that could bear upon the NEO debate *inter alia*. Lately, of course, the 'fake news' phenomenon has been causing especial concern.

As regards the Gildas depiction, Victor Clube and Bill Napier were disposed to see the background reality as astronomical, a meteor stream or fragmented comet the British Isles were passing through.[18] The migration of thousands of people from the south of Britain to what, by the mid-fifth century, was coming to be identified as "Brittany" indeed seems to point to some such sudden trauma affecting Britain. After all, these migrants came in spite of the Huns still posing a threat to the whole of Gaul. This they were to do until their defeat in 451 at the battle of Catalaunian Fields near Troyes.

It seems, too, that astronomical causation gains traction from traditions which tell of the death then of Vortigern (the Romano-British leader who had invited the Saxons over) along with the demise of a leadership cadre, just possibly his confederal council in session. There is, too, an ominous entry in the *Annales Cambriae*, a contemporary compilation. It tells of "darkened skies" in 444 or thereabouts; and this may well refer to a dust cloud over a very extended area for perhaps several months. Likewise Gildas does speak of a "dreadful and notorious" famine that suddenly capped a food shortage brought on by armed strife. He indicated this was raging around the time of a plea for assistance made to Actius, the Roman commander in Gaul. This will have been in 446 or thereabouts. A Gallic commentator had noted in 442 that an English offensive had got under way.[19] That same year, the Chinese had registered "a new comet". May it actually have been a distant meteor stream?

The sequence overall may have been something like this. From 442 or 443, there were several years of exceptionally heavy meteor incidence. The wider ecological impact was northern hemispheric if not worldwide. Witness treeline recession in Scandinavia and Canada. Apparently some of the scars were still evident a century later, notably in the forests and the towns.

Still, this does not mean that abnormal meteoric activity caused Romano-Britain simply to collapse in the face of English pressure. The contrary can certainly be seen in how, around the turn of the sixth century (i.e. *c.* 600 AD), British cavalry were to inflict a severe defeat on the English at the Battle of Mons Badonicus. Opinions range widely about where this might have been. My own disposition, longstanding but yet to be properly tested, has been to locate it about the South

Chilterns escarpment – that is to say, it was an offensive sweep progressing well to eastwards.

As regards Gildas, there has recently been a tendency to depict him as just rhetorical and confused.[20] However, if the meteoric shower is fed in, he comes across as earnest but reasonable, much as he seems to have to his contemporaries.

Meteoric influence, a longer view

In 1966, the journal *Acta Astronomica Sinica* noted 147 major meteor showers being observed over China since 1575 BC. Ten were recorded in the fifth century AD: two in 443 and 447 with all but one of the remaining eight being between then and 466. This pattern seems to relate well enough to the interpretation just rendered.

It may do so the more in that the 443 to 466 span represents a climax in this regard across eight or more centuries. Thus the sequel across the AD 100 to 1100 time frame is as follows. No major showers are said to have been observed in the second century; two in the third; one in the fourth; those ten mentioned above in the fifth; five in the sixth; nil in the seventh; six in the eighth; nine in the ninth; eleven in the tenth; and six in the eleventh.[21]

To Here From Eternity

There are other galactic phenomena to consider. Large but very diffuse clouds may occasionally be encountered in interstellar Space, these mainly comprised of hydrogen. Clube and Napier also drew attention to our Milky Way being one of the many spiral galaxies; and to our Sun putatively having resided within one of its spiral arms for tens of millions of years.[22] However, in regard to spiral galactic circulation overall, Theodore Snow has submitted that "the density of stars between the arms is nearly the same as it is within one".[23]

10

Celestial Engagement

(1) Strategic Command

The precept guiding this study has been that the five permanent members of the UN Security Council should be encouraged to lead humankind towards the creation of a Cosmic Command staffed by volunteers that, maybe down to battalion level, are seamlessly integrated, racially and nationality-wise. As the title suggested implies, the proposed command would be intended, in the first instance, for defence against incoming NEOs. But protection against similar terrestrial disasters (supervolcanoes, mega-earthquakes, pandemics . . .) should in due course be brought within its remit. English would perforce be the operational language, at least against all threats from Space.

The command's key resources and procedures would have not to be an assemblage which could readily be used in a putsch to establish an authoritarian world government. See Arnold Toynbee and Bertrand Russell as quoted in Chapter 6. Most importantly, its only crew-served weapons should be ones designed very specifically for use in NEO interceptions. No doubt, too, there would have to be a Cosmic Supreme Commander, presumably appointed by the UN Security Council. He or she might best be a citizen of the United States or some other permanent member. Their deputy not so? Terrestrial national forces, organized mainly for regular war, would respectively be kept at some countervailing strength too.

Contrary Indications

Not that any such matrix of checks and balances would be easy to create and maintain. Granted, no international consortium or single country can legally sustain a sovereign claim to any part of Near Space, outside the vertical limit so informally set for national air spaces. However, influential Pentagon opinion has already been questing "full spectrum Space control", meaning an overall American military dominance of Near Space sufficient to ensure unilaterally that whatever national Space policy the United States wishes to pursue cannot be thwarted by any other party, forcefully acting within the Space realm. Other national programmes or individuals from them would still be welcome to join particular projects but this would be *ad hoc* on an essentially subordinate basis. All else apart, there has been a longstanding belief that full collaboration with non-Americans tends to raise the total cost of any such exercise by up to a third.[1] But such received reason would needs be reconsidered within the context of a Cosmic Command.

Supranational Authority

This Command could face vexing problems in respect of deciding when one should open fire on an intruding NEO, be this with nuclear fission or thermonuclear warheads or otherwise. What ought to be the procedure for devolving authorization from the UN Security Council? Broadly similar dilemmas had to be faced by NATO over sixty years ago apropos theatre nuclear weapons.

The answers the alliance eventually firmed up on were understood to have been, in general terms, that authorization would have to be granted by the President of the United States and by the government of any country from which the said weapons might be fired, along with that or those of the forces operating the particular systems designated. The cosmic defence strategy here being propounded could admit of nuclear authorization devolving, for the near term, from the permanent members of the Security Council, presumably acting on a unanimity basis and using guidelines hopefully agreed well in advance.

The operational soundness of any command-and-control procedure is going to be determined largely by the amount of time the Defence has to prepare a response. A significant number of NEOs have recently come at least fairly close to the Earth, often being detected late. In February 2013, for instance, a superbolide (i.e. an exceptionally large meteorite) was not discerned professionally at all before it airbursted low over Chelyabinsk, Russia, causing 1500 casualties albeit no fatalities. Now the big advances being made in the quality, variety and density of telescopic coverage should soon mean that (a) very few asteroids or whatever will come close undetected, and (b) asteroids or comets on more or less elliptical courses which are potentially threatening may sometimes be tracked decades or even centuries ahead of prospective impact, maybe many orbits later.

In this invaluable respect astral defence will enjoy much better prospects than NATO cross-border defence did pre-1990. All in all, one can say with high confidence that action decisions affecting the outlook for life on Earth will rarely have to be taken within a few hours. This in itself will be a boon for any cosmic military strategy.

But background questions will be on-going regardless. Many will concern the design parameters of any interceptor missiles and their support facilities. What speeds, manoeuvrability, ranges, warloads and electronic support should be sought? And how far should a Cosmic Command go in working to a mix of specifications? Should it be looking, among other things, towards a few rockets that can travel outwards in excess of, say, 50,000 miles? Should there for early use be ones with nuclear fission warheads? Should not thermonuclear megaton-range warheads be available if reconnaissance warranted? What might we one day need in terms of surveillance and/or interception facilities on the Moon, Mars or modular platforms in orbit? In short, how should the overall architecture evolve?

A precedent that invites comparison is the flurried debate that broke out in the late fifties within NATO but especially in the United States over the introduction of battlefield nuclear weapons. Edward Teller tended to favour some in-theatre reliance on them as especially suited, in his view, to the adaptive military cultures of the West. Henry Kissinger thought somewhat similarly. But Robert McNamara, US Secretary of Defense from 1961 to 1969, was not persuaded. Late in 1960, an order had provisionally been placed by the Pentagon for 6250 Davy Crockett "atomic mortars", each one able to fire a shortish distance a nuclear warhead yielding the energy equivalent of about a quarter of a kiloton of TNT. The order was not proceeded with.

Nevertheless, the American distribution within NATO of several thousand battlefield nuclear warheads had gone ahead, some of the total being for American forces but most for NATO allies. At the end of 1963, the warheads then in this Cold War stockpile were said by Secretary McNamara to have an explosive yield of 400 megatons.[2] Well before then, however, he had told President Kennedy that he did not believe battlefield nuclear weapons could be used to fight a European war to a positive conclusion. In other words, the alliance had arrived at a strategy which secured allied compliance in peacetime rather than assured survival in a warlike crisis. A future Cosmic Command would need to do better.

The point is only too well confirmed by how a sizeable fraction of this Cold War total megatonnage was assigned to NATO air arms operating off long and vulnerable runways on their European bases.

These planes were understood to be a key part of a panoply of nuclear power intended to inhibit escalation. You could say rather that they veritably invited it. Moreover, as the USSR was soon to be enveloped in a final crisis that might well not have ended as peaceably as it did, this inference became more than mere shadow boxing. During these years, this author put the point in discussion with the then Supreme Allied Commander Europe (SACEUR). His gracious rejoinder was that "this was how things had evolved historically". Of what circumstances could this not be said?

The Fourth Domain?

The notion of Near Space as a fourth domain in war is interwoven closely with the concept of it as a zone without war, without terrestrial war at any rate. All else apart, sustained terrestrial conflict out there, involving much expenditure of hardware (especially missiles and orbital satellites), would rapidly make debris loading in Space quite unmanageable. Just about the only pattern of Near Space conflict that might initially have a slither of reason attaching to it would be the non-destructive jamming of electronic signals within small segments of Near Space. Yet the logic of even that scenario would surely dictate that the said conflict would have to either cease swiftly or escalate markedly. Best simply to accept that the counter-culture people have a point when they say that "the heavens were made for wonder, not for war". Acceptance thereof might well help to dampen down armed confrontations nearer the surface.

(2) Operational Applications

How specialists in the anti-Near Earth Objects field share knowledge with their peoples has evolved the last two or three decades. Even so, one cannot but admit to some mystification at an apparent tendency still to discount in discourse the constraints imposed on the defence by the "velocity of light", a value constant for the several electromagnetic radiant manifestations: light, heat, radio, radar. Thus for a radar pulse to travel ten million miles and back will take close to two minutes. During that time, a comet could have travelled thousands of miles. And it might be swinging rhythmically the while.As regards the actual mode of destructive interception, two basic alternatives continue to present themselves. The one is to seek to divert the NEO from its menacing path. The other is to reduce the said object to smithereens. Again, the former choice can be divided in two. The one option may be to ease it off its present course by gravitational attraction. The other is to do so by dynamic physical contact.

On first sight, the "reduction to smithereens" approach looks like the one to opt for. However, the complication remains that the end product might prove to be not smithereens but those largish fragments so scattered as to be liable to present collectively a bigger threat to the Earth below than the unitary NEO would have done. Such could easily turn out to be the case if (a) the interceptor missile bore a proximity fuse primed to activate its warhead within a very narrow distance band, in front of or behind the NEO, or else (b) had embedded the warhead in the NEO, this while registering a closing speed of perhaps tens of miles a second.[3]

In 2005, Edward Lu and Stanley Love introduced into the debate the alternative concept of using "gravity as a towline". Their reckoning was that a 20-ton nuclear-electric Spacecraft hovering 50 metres or so above an NEO for perhaps a year could deflect a typical asteroid (say, one some 200 metres across) sufficiently to ensure that, within ten or twenty years, it would no longer be liable to hit the Earth.[4] At the Jet Propulsion Laboratory at Pasadena, Donald Yeomans has been disposed to see this technique as mainly proffering scope for "trim manoeuvres".[5] But much the same might be said of the physical nudging of orbital NEOs in order to preclude their collision with the Earth perhaps some orbits ahead. Again, early warning times may thus be measurable in months or years.

It is clear that, for optimum planning and execution of a given mitigation attempt, a reconnaissance Space mission (manned or unmanned) will often be mandatory.[6] This particularly applies to the larger asteroids as opposed to the comets because the former are structurally and materially more complex, therefore individually less predictable in relation to such aspects as shape change, solidity and spin rates. Asteroid albedos can vary by a factor of five; and densities do so three-fold, this latter spread being quite closely related to porosity differences. Coloration can also be mixed. All in all, "monster asteroids are far more dangerous than comets. The latter are relatively soft and, of course, there are far fewer of them".[7] In either NEO category, though especially the comets, "monster" could sometimes mean a mass of a million tons.

Anti-NEO Nuclear Explosions

Nuclear explosions up to the low megaton range are widely discussed as a prospective part of the anti-NEO panoply. Thus, "For larger objects and/or shorter warning times, impractical numbers of kinetic impactors may be necessary for a satisfactory deflection. In that case, nuclear explosions are an alternative . . . capabilities exist for dependable yields from about 100 tons of high explosive equivalent to a megaton or more"[8]: in other words, into the thermonuclear range.

Donald Yeomans has further submitted that the most effective kind of explosion in this context would be a fusion reaction "that could generate copious quantities of neutrons that would penetrate the asteroid's surface, strongly heat and vaporize the surface material and provide the recoil push to alter the asteroid's trajectory".[9]

He has also noted that David Dearborn at Lawrence Livermore has done computer simulations indicating *inter alia* that a 300-kiloton explosion from a charge embedded a few metres into the surface of an asteroid 270 metres across, say, might so fragment and dispose of its mass that only a few per cent thereof would hit the Earth "even if the explosion occurred but a few weeks prior to the expected impact".[10]

The 1963 Partial Test Ban Treaty and the 1996 Comprehensive Test Ban Treaty prohibit nuclear warheads in Space. Together these enactments could constitute a serious obstacle to anti-NEO preparations by a single state or several. But the treaties being rescinded or modified within a United Nations setting might well allow of the creation of a viable planetary defence screen. For recourse to fission or fusion warheads against NEOs as necessary could be a key part of the answer. With it might come a commitment not to use fusion warheads in regular terrestrial war unless the adversary was already so doing; and to use fission warheads in such a conflict only upon a due political decision being taken. Use either kind of nuclear warhead against a biological attack provided this was adjudged politically to be an imperative response to a threat one felt confident had emanated from an identified source? Throughout, the accent should be, as now, on minimalism.

Nuclear warheads exploded thousands of miles in Space will not produce a heat flash injurious to life on Earth. The radioactive effects should be avoidable as well. Nor do we have to assume that the explosive use of nuclear energy at a great distance into Space would harden attitudes in favour of its use near ourselves and other terrestrial life. My own view is that it would shift opinion more the other way.

Nor should one be dissuaded by the notion that one should never use nuclear warheads at all, as a matter of principle. Granted, a case can readily be made in terms of the macabre genetic effects. But suppose the Allies had developed nuclear warheads by 1943, say, instead of 1945. Would it have been wrong to use them to stop the Holocaust and the plethora of collateral fascist atrocities?

11

Emergent Priorities

The new world we are moulding, however imperfectly, could turn out, by 2150 or so, to be either a sight better for most people or else a good deal worse. This divergence will stem in part from "acts of God", terrestrial or cosmic. But "worse" could otherwise be due to our persistent lack of integrated foresight in government. All else apart, this is what President Trump's USA will be anxiously judged on.

Open Conspiracy

Between the two World Wars, H. G. Wells envisioned an "open conspiracy" whereby the spread of awareness and knowledge, informally and otherwise, would bring more and more people everywhere to insist on "the provisional nature of existing governments and of our acquiescence in them" and point towards economic and social management by, so he vacuously presumed, a "world directorate". [1]

New ways of thinking, which transcend existing boundaries (social, political, national, religious, scientific, pedagogical . . .) are needed to cope with the situation on-coming. The notion that such a "conspiracy" might henceforth be held together not by Esperanto penpals but via satellites in orbit is said to have originated around 1945. Credit is accorded Arthur C. Clarke, not an "HG" then deeply melancholic in terminal decline.

Space Odysseys?

An emergent commonality is the professional interest being shown in whatever prospects there might be for eventually "conquering" Space by dint of having adventurers visit destinations outside our solar system. Not that the dawn of this day can be at all close, whatever the reckoning. After all, the nearest star beyond our Sun is over four light years away. Granted, it does seem that powerful laser beams might one day impel a Space craft to attain a significant minor fraction (20 per cent?) of the speed of light. Given the inevitability of beam divergence, however subtle, co-ordinated batteries of lasers hundreds strong might be needed to achieve results over extended ranges.

On 12 April 2016 (the 55th anniversary of Yuri Gagarin's death), a Russian physicist and billionaire by name of Yuri Milner announced his intention of developing interstellar flight, manned or unmanned.[2] His plans have been unfolding since. His aspiration looks wildly optimistic. But it is at least a sign that Russia's post-Soviet turn off from Space is gradually on the wane.

Further exploration of the solar system, manned and unmanned, has lately been envisaged as a follow-through of the era which reached its climax with the conclusion in 1977 of the four American planetary fly-bys using the unmanned Spacecraft, Voyager 1 and 2. Manned flights to emulate these solar system missions are now being actively prepared for. So are unmanned interstellar probes.[3]

It already seems apparent (notably from distinctive absorption lines in stellar radiation) that the physics of our galactic group and well beyond is subject to the same laws throughout. Yet across the years, not a few people have been exercised by the thought that this may not be quite the case in exobiology.[4]

An International Institute for Planetary Studies?

To my mind, there is a clutch of reasons for believing that skyward understanding would be enhanced if an International Institute for Cosmic Studies (IICS) were to be established, maybe along IISS lines. A rich mix of policy and science issues would come up, more or less regardless of how world affairs were unfolding otherwise. All else apart, it would be a good think tank for Cosmic Command, inspirational as well as informative.

Many of the particular issues would be interdisciplinary. Interplay between governments and academe would wax important. So would that between specialists and generalists, between specialists and any "open conspiracy".

Multinational Regionalism

Otherwise our world emergent will continue to be characterized by an on-going proliferation of formal linkages between countries within the same geographical region or whatever. Tariff reductions and other economic measures have figured especially. Since 1994, there have averagely been worldwide ten regional deals per year though only the one comparable global deal, the World Trade Organization's "underwhelming" Bali package of 2013. Additionally, regional bodies either are or might become bulwarks of stability in realms largely or wholely outside Economics as formally defined. One thinks of the control of insurgency, urban crime, pandemics, nuclear management, rural ecology, scholarly enquiry, natural catastrophes. The respective boundaries do not have to coincide. It may be healthier if they do not.

May one take a cautionary look at two world regionalisms among the most remarked in regard to wider purposes – the European Union (EU), originating as the Common Market in 1958; and the Association of South-East Asian Nations (ASEAN), formed in 1967 in anticipation of substantial disengagement from Far Eastern security on the part of Britain and the USA. The number of member states has grown over time in each: from five to ten in the case of ASEAN and from six to 28 (with several more now aspirant) within the EU. At the moment, thinking about the latter is dominated by "Brexit".

Brexit

May one now simply focus on several issues which were flagrantly neglected in the course of a 2016 "Brexit" referendum debate which was to prove, in any case, too ill-informed and insubstantial. One would be whether the Middle Eastern and African countries (from which so many migrants are now arbitrarily seeking to enter the EU) are not thereby losing far too many of their most talented, enterprising and qualified people.

The obverse side thereof was how the once so beautiful South-East of England, in particular, could become critically overcrowded by migration from outer Britain but also from across the English Channel. The very real pleasure and stimulus which might have been proffered by a more moderate trend to communal diversity could thus be compromised fatefully. We cannot just assume that any adverse tendency can readily be corrected as per the truly harmonious relations now prevalent between the wave of tropical Commonwealth migration to Britain from 1960 or thereabouts and the indigenous population. This mood took decades to engender.

Turkey and Europe

The outlook in this regard has been further complicated of late by a still lingering prospect of Turkey's accession to "Europe". Kemal Ataturk, having become his country's first President in 1923, committed her to secularization interwoven with progressive democracy. And until recently at least, another cardinal principle, "revolutionism" involved an unyielding military preparedness to act forcefully against reactionary backlashes.

However, with the savage governmental crushing of the 2016 military coup, Turkish politics has reverted to being much more aggressively Islamic. Moreover, a serious omission from 1923 onwards has been the absence of a master plan for revitalizing her traditionalist countryside. As of 2008, the percentage of the national workforce engaged in agriculture was 26 still as against between one and three in Australia, Britain, France, Germany or the USA.[5] Agricultural modernization under the auspices of the EU would eventually release well over ten million farming people for livelihoods not dependent on the land. Britain could once again be, for the usual linguistic reason, a prime migration target.

Meanwhile, the "Britain and Europe" question is engendering a great sense of crisis on all sides. At its heart is a broad desire to see Britain play a strong European role still, this alongside the European Union. "Europe" could thereby become, more than she is already, a regionalism without the one tightly defined boundary appertaining to all purposes. It might thereby be a model for an international order not set so hard and fast as to embark readily on destructive conflicts.

May one add this. Society more or less worldwide is going to change the next few decades at a pace quite unprecedented and in more directions than ever before. Above all, people will generally be living progressively longer; and, thanks to automation, working considerably less. And may not birth rates revive somewhat? See below.

Governments in Europe as elsewhere will be likely to respond to a changing situation in contrasting ways. The contrasts will surely put a premium on controlling migration across national boundaries within the EU. So could we not have this by 2030? And would this not weaken the case for a very categoric Brexit?

The Mekong Crisis

The evolution of ASEAN was fairly fully explored in *The Bounds of Liberalism.*[6] Suffice now to stress that the Mekong river basin, which serves so vitally its core area, is at serious risk of collapsing as an ecological and economic asset. At 795,000 square kilometres, the basin comprises a large portion of Vietnam's farmland as well as of Cambodia, Laos, Thailand, Myanmar and the Yunnan province in

China. At present, there is often a ratio of 30 or more to one as between critical water flows in the respective wet and dry seasons; and such disparity can be expected to become more prevalent as winter snow melts more readily in Tibet. Temperatures in the Mekong delta rose 0.5°C between 1979 and 2009; and are expected, by Vietnamese climatologists, to rise further (somewhere between 1.1 and 3.6°C) by 2100. More severe tropical storms are also anticipated with this climate change.

Unfortunately, the number of states involved and the complex mix of managerial variables encourage secrecy, short-term thinking and generalized expediency apropos Mekong water control. In 2009, an expert opinion on this syndrome was that "none of the current proposals or projects for mainstream dams could pass even the most cursory environmental impact assessment".[7] A 2015 informed opinion has been that "for the sake of a little more industrial power, the gifts of the river are being squandered. Life as it has been is not life as it will be. The days of stepping into the river are numbered".[8] And beyond the Mekong, we are still left with Indonesia being so much bigger than any other ASEAN member.

Qualitative Underemployment

Leading on from international regionalism, this author has not yet seen, as regards social change, a great deal to add to or subtract from his recent trilogy of Sussex Academic Press studies. But certain tenets should be revisited.

One that clearly calls for added emphasis is "qualitative underemployment". Loss of fulfilment may be incurred through the relinquishing of distinctive skills and singular experience in favour of the commonplace routine that so often comes with automation. Correspondingly, less value may attach to human qualities over and above mere functional proficiency. One thinks of physical bravery, fortitude, integrity, courage, empathy, leadership . . .

Malthus Refined

Also to reckon with, however, may be a tendency for the population explosion to slide out of control again globally, albeit with the economically more developed societies making the running this time round. Occasional pandemics excepted, life expectancy looks set to increase incrementally for many years ahead. Correspondingly, the profile of the quality of life medically speaking should extend, in some respects, nearly in phase with extending longevity.

Renewed demographic expansion could stem, too, from a desire to rebuild the "extended family" concept in order to compensate for a weakening of local community identities. It could also help to offset reduced fulfilment in the workplace, as and where there is a workplace to be unfulfilled in.

Public Expenditure

Needless to say, adjustment to these interrelated developments will come to involve heavy increments in public expenditure. These will especially be on health and education throughout life. So may they be on public pensions throughout adulthood, to be spent as individuals wish. But urban renewal in the round, in a quest for community identities and leisure requirements, needs also to be factored in. And by 2030 the wider distancing likely deemed imperative between driverless vehicles in motion may be compounding markedly urban traffic congestion. More leisure will also mean more pressure for urban dwelling space. No doubt, too, there will have to be substantial outlays on physical protection against international aggression, insurgency, apolitical crime, natural disasters and other episodic occurrences.

War and Rumours Thereof

Implicit in what has been said about the safeguarding of peoples, collectively and individually, is the proposition that this will be the sector of public expenditure in which it has become easiest to get things fundamentally wrong. The pace of change means big mistakes will often be made in educational philosophy, health care or whatever. But they may usually be corrected with more assurance than can be the case with defence postures, cosmic defence included.

Military Dispositions

Unfortunately states, especially authoritarian ones, have often seemed only too ready to maintain armed forces that are ill shapen and overly large in relation to legitimate interests and aims. This they may do because of a fear of internal upheaval. They may do so, too, because their military have been left free to act as a vested interest. Alternatively or additionally, they may act thus because a strong military presence may exhibit a martial puritanism as an alternative to a distractive consumerism which could otherwise arise, inducing markedly exotic life styles and aspirations.

Soviet Military Precedence

We might take, by way of illustration, the former USSR, focussing on the naval and strategic air postures of this mighty continental polity. Through the 1950s, the Soviet Navy acquired no fewer than 600 submarines, nearly twice the number Nazi Germany had maximally possessed. No matter that her planners can hardly have overlooked how many of these boats were too small for Atlantic or Pacific patrols. Nor that Soviet operational doctrine said a general war would go all-out nuclear, immediately or very soon. Once this had happened, the critical Allied marine problem would have been not a shortage of ships but a dearth of usable ports for them to ply between.

Similarly, in 1990 the Soviet Navy also retained in service no fewer than 350 minesweepers, many of them designed for inshore or coastal service against a mine threat hardly worth presenting in lonely waters so landlocked or ice-locked.

Moderation

The precept that we should, in any case, try to lay the foundations for an age in which the noblest form of conflict will be that against Near Earth Objects (NEOs) puts a premium on pursuing any terrestrial conflict only to a point at which essential aims can be realized. A pertinent commentary on military affairs is afforded us by Sun-tsu, a contemporary and compatriot of Confucius (551–479 BC) and the first military strategic theorist known to History. The first and arguably the greatest. Like Confucius, Sun-tsu never doubted the virtues of peace. Like him, too, the latter stressed psychology in his pursuit of it. Witness the averration in his famous text, *The Art of War*, that it is better "to take intact a regiment or company than to destroy it". What is of supreme importance in war is "to attack the enemy's strategy".[9] Embracing Geography with some emphasis, he stressed the importance of "focal ground". Thus do not fight with your back to a waterline. Get across to the other side.

Throughout this early work there is a sense of equivalence between opposing armies. No hint of triumphalism, whether one is looking forwards or back. No scorn for nomadic tribesmen.

Disengagement

In the present climate of global opinion, however, endeavours to create a Cosmic Command or anything broadly akin, seem likely to meet with not a little studied indifference. For there has lately been, across the West and beyond, a spontaneous but widely cast reaction against the whole political process, democratic and otherwise. At the heart of it is a simmering sense that acting as a representative is a most unrepresentative thing to do.

How long will this mood last? A decade or two? While it is extant, it will engender scepticism as to whether anti-NEO defence can ever be made to protect terrestrial life sufficiently while serving also to spearhead a posture defining anew our relationship with the cosmos at large.

The case has herein been put with emphasis on any on-going cosmic human presence being drawn quite tightly within our solar system. This would be in keeping with our being a species given to territorial definition. Besides which, we would not wish to have our fellows spend several decades on a round trip to a nearby star only to find, in all likelihood, nothing to add to our present knowledge, except perhaps menacing microbes.

Lunar Early Warning?

Close within our solar system, the balance of costs and benefits could look more favourable throughout. Not least might this be as regards our Moon, a celestial body almost certainly microbe free. A lunar manned observatory would have pronounced operational advantages over any terrestrial one. Witness the near absence of local electromagnetic interference; no atmospheric absorption; no debris in suspension; and little overlap of return signals, interference from which is worsening rapidly on Earth. Needless to say, low interference could help make lunar facilities critical for the detection, tracking and interception of NEOs. But a caveat to enter, in operational terms, is that facilities on the airless Moon would be exposed to bombardment by meteorites. So, of course, could be any independent modular platforms deployed by Cosmic Command out in Near Space for early warning or engagement.

The Moon and Mars

Then again, the Moon could be made a short-trip tourist centre several decades hence. But this does raise doubts again about whether we ought to lay on excursions so demanding of resources at a time when we shall still be seeking ecological and economic stabilization on Earth, this on the basis of commonality of resource restraint.

Yet that leads on to the overarching question of whether the Moon should be put to any mundane purposes, given the part we know it has played for thousands of years in relation to the spiritual sensibilities of humankind. A short overview of this dialectic is presented in *Engaging the Cosmos.*[10] Suffice to note for the moment that what look like rudimentary lunar calendars have been found from *c.* 20,000 BP.

Apropos Mars, a further sorting of priorities will have to take place. Science fiction writer, Brian Aldiss, has reportedly argued that we should think of Mars as an "Ayers Rock in the sky. Mars . . . has a grandeur and appeal of its own. Mars and the other worlds have come to offer us more than crude physical resources; they are sources for the spirit as well."[11,12] So no anti-NEO installations out there at all? Or just the one complex? Then again, one cosmic idiosyncrasy can be said to give especial expression to the non-sectarian spirituality the "heavens above" can still be said to invoke. It is the irregular alternation, as we see it, of the Moon and the Sun. The former "at the full" looks almost exactly the same size in the heavens as does the latter.

An Anti-NEO Canopy?

A mutually perceived need to latch into each other's resources to form a canopy against NEOs could draw together to positive effect polities in contention on Earth. But the operational results may still be mixed. A report one has perused of the tracking of an asteroid some distance out indicated that its positioning at a given point in time had characteristically been misjudged by 12 to 15 kilometres, a discrepancy apparently taken as broadly acceptable yet surely too negative, certainly in terms of non-explosive nudgings.

Likewise a large comet or asteroid with a mass of maybe several hundred thousand tons will be very hard to impact to effect. This will especially apply when a proximity fuse fitted to the interceptor is being used to trigger the disintegration of an NEO comet, say, at a required distance. Given that the speed at which a rising interceptor is closing with an incoming NEO head on can be close to 100 kilometres a second, there may be only the briefest of timing discrepancies between the NEO's largely disintegrating on contact into a mass of harmless particles and its simply breaking into several menacing fragments. Having said that, an oblique impact leading to a partial loss of material may affect decisively NEO speed and path changes. However, the fireball from a one megaton explosion might take some twenty seconds to achieve a diameter of, say, six kilometres.[13] During that interval, the NEO might travel several hundred.

Then again, deflections by gravity traction might be hard to achieve, given too massive and mobile an NEO. Spraying the hostile body to effect route change through differential heat absorption is a recognized option too. But how readily and how potently?

A provisional inference may be that the best hope for NEO interception will be when effecting the non-explosive nudging of asteroids repeatedly orbiting the Sun and coming a little closer to the Earth each time round. Otherwise, the emphasis may have often to be on passive defence, the emergency evacuation of a specific land area and/or any coast thus exposed to a tsunami.

What can still be looked towards, however, is a gradual improvement across the board in the tracking and acquisition of NEOs, over and beyond the comparative advantage of one technique or one kind of target as against others.

Optimism on this score used to be upheld in terms of the precept Gordon Moore enunciated in 1965. This was that the characteristic number of transistors on a chip would continue to double every couple of years. However, "Moore's law" has all but run its course because transistors have come down in size to an atomic scale. In 1965, a few hundred transistors might be "crammed onto a slither of silicon. Today's powerful chips contain billions." But other design developments are keeping the efficiency gains going.[14] And anti-NEO defences gradually getting denser all round should also become a positive aspect. There could eventually be scores of anti-NEO batteries on the Earth's land and sea or elsewhere in Near Space.

12

The Cosmic Setting

The last quarter of a century or so, professional awareness of our planet's exposure to impact by Near Earth Objects (NEOs) has become more acute and institutionalized. In 1992, a US Congressional study group produced a Spaceguard Survey report which presented NASA with a mandate to record, within the next ten years, 90 per cent of all locatable NEOs a kilometre or more in main diameter. Then in 2005, a Near Earth Object Survey Act was ratified by Congress. It tasked NASA with locating NEO asteroids not less than 140 metres in diameter; and with reporting to Congress annually on the progress made. By 2014, the lag of progress behind expectations was actually getting wider. But this was primarily because large radio telescopes were showing sizeable NEOs to be a deal more numerous than expected.

The universe we can actually feel we know something of appears "isotropic" – that is to say, it is apparently physically structured throughout in accordance with the same principles and material composition. As much could be said of its biology, save that evolution is prone to take many alternative tracks. Any alien life, however encountered, would almost certainly be microbial in whole or large part. Whether its larger structures are consistently cellular microscopically remains uncertain. Whether alien direct contact would lead to infection, in either direction, is guesswork. But there has to be a risk that it would.

For planetary geopolitics to be appropriately benign, it will be important that any anti-NEO screen can be so shaped and managed that it cannot readily be redirected against targets on the Earth's surface. Meeting this requirement will be made the easier if at least part of the defence screen can usefully be designed to intercept tens of thousands of kilometres out. A rationale for such a stipulation might be that an NEO may be too liable to break into those largish fragments which we would then need to have dispersed well clear of the Earth's silhouette before they can arrive in its general vicinity.

Unmasked by Science

If you ask your archetypal commuter what he or she thinks of the cosmos, a representative answer might well be that they are dumbfounded by the scale of it all. Not infrequently, indeed, the same has been said by astral cognoscenti. Thus in discussion of early modern French thought, Blaise Pascal (1623–1662) is regularly counterpoised with René Descartes. Often considered hardly less brilliant as a mathematician, the former came to feel that rationalism – Cartesian or whatever – is not enough. For many, his most enduring line, "The eternal silence of these infinite spaces frightens me", showed him to be "not one man but several". His French anti-papal Catholicism allegedly meant that he "never achieved mental unity".[1] A complacent Voltaire was more than happy to dismiss Pascal as but a melancholy misanthrope.

During the "interwar" period, James Jeans afforded British astronomy some inspirational leadership. However, he admitted to having been overawed by Creation as he first perceived it but claimed to have more or less come to terms with everything by dint of its majestic beauty. A generation earlier we find equivocation of a different mien emanating from Jules Verne (1828–1905). He began his literary career with plays and short stories, these often deeply critical of Science. But in due course, he found a publisher, Pierre Herzel, who was insistent on his writing positive and duly lucrative science fiction. Then after Herzel's death in 1886, Verne reverted to his erstwhile concerns, championing environmentalism, anti-capitalism, and social responsibility, questioning the while "the benefits science and technology could bring to an imperfect world".[2]

String Theory

Still, a big challenge today does not lie in differences in philosophical perspective. Nor does it in how counter-intuitive cosmological concepts and data can sometimes seem. It resides rather in the "string theory" requirement to introduce extra dimensions to our molecular reasoning, these to accommodate the equations needed to resolve basic analysis throughout Science. The said stipulation is usually traced back to 1919 when Theodor Kalusa backed up by Oscar Klein, both Swedish, proposed an extension of the Space–Time relativity dimensions from four (time plus three spatial) to five in order to incorporate electromagnetism. The fifth was depicted as being "curled up" in a tiny heap, a mode of which our sensory attributes would be unaware.

Within a few years, the Kalusa-Klein reasoning had fallen somewhat into abeyance, the science community at large feeling ill at ease about the added dimension. However, around 1980, string theory was revived and extended to comprehend ten dimensions.

Lately, "string theory" thought experiments about the fundamental nature of the material world have been pointing towards the ultimate possibility of 10^{500} quite distinct universes, 10 with 499 zeros after it before you reach the decimal point. By "distinct" one here can mean operating within dimensions utterly inaccessible from the outside.

If instead one should still prefer to believe our universe stands complete as an infinite unity, one is up against Olber's Paradox. Some two hundred years ago, the Bremen physician and astronomer, Heinrich Olbers, emphasized that if our universe stood alone as boundless Creation, eternal and unending, limitless light would be flooding across it from every direction. Both observation and plain common sense worked against such envisaging.[3]

Religious Inclusivity?

Against our evolving perspectives in Philosophy, Cosmology and the Life Sciences, interest is being shown within the major faiths in there being more positive interaction between their respective obediences as well as between them and Science. Thus the present (and maybe the last) Dalai Lama, Tenzin Gyatso, was stressing come the turn of the century how such convergence could enhance true spirituality: love, compassion, patience, toleration, forgiveness, humility, appreciation of Nature . . .

An irreligious person may, the Dalai Lama has felt, find the pursuance of all this much harder to sustain. He has suggested joint pilgrimages as among the possible modes of inter-faith dialogue; and has proposed that the end game could be "a genuine sense of religious pluralism", an outlook hopefully able to avoid both bigotry and "unnecessary syncreticism".[4]

Not long afterwards came a quest for sectarian accommodation which bore directly on the fraught Middle East. British Jewry was recently blessed with a Chief Rabbi in the person of Jonathan Sacks, someone possessed of strong liberality.[5] In 2002, he completed *The Dignity of Difference: How to avoid the clash of civilisations.* He stressed how every great faith has texts which, read literally, "enclose narrow particularisms, suspicion of strangers and intolerance towards those who believe differently. Every great faith also has within it sources that emphasize kinship with the stranger".[6]

Correspondingly, he called for an Israeli–Palestinian peace based on a two-state division of the Holy Land, this to be closely coincident with demographic variation. It would need also to include an agreement on Jerusalem and its Holy Places; joint supervision of water and other shared resources; and an international accord about refugees.[7]

May one now propose, too, that further to any such dispensation, dual nationality citizenship could hold the key to Jerusalem being united on the basis of democratic choice. Furthermore, if a genuine peace based on two-state or confederal democracy is not built in the Holy Land, the Middle East (plagued by rainfall decline to add to its other ills) could lead the world into an era of planetary fascism (blue, red or hybrid) hardly less hate-laden than what emerged post-1918. For the massive information explosion now under way means that polities everywhere are destined to become either considerably more liberal or else a lot more oppressive.

Cultural Flux

The kind of ecumenism envisaged by the Dalai Lama and Rabbi Sacks may well have a lead part to play in bringing our planet nearer to a long enduring peace. But, with respect, one has to say that, in themselves, their renderings were insufficient. The principle of togetherness has further to extend across the adherents of a wide variety of value systems – national, linguistic, military, nuclear, cosmological, ecological, agnostic, humanist, handicrafts, music, fine arts. . . . As the different patterns overlap within social groups and in the makeup of individuals, the particularities will either blend or diverge.

To my parents' generation, to ask somebody if they believed either in God or in human immortality was to put the same question in alternative forms. Now it is becoming fairly normal to express belief in a kind of godliness, a creation principle if you will, without understanding this to encompass necessarily the notion of individual human survival after death. Astronomy is playing a large part in this.

Material Culture

Meanwhile, the more material aspects of culture change are poised to pose searching questions about individual fulfilment and hence about planetary peace. And because there is an unprecedented pace of cultural change under way in many directions, the challenges thereby presented are acute.

As one writes, NASA is rightfully celebrating the achievement of having an unmanned Space probe orbit Jupiter. What is thereby discovered may give us helpful indications concerning the form and frequency of planetary systems round other stars. It could also prove to be an encouraging case study in long-range telecommunications, something this text has expressed concern about more generally. But how much more exploration of a similar kind can usefully follow?

It is salutary to recall what Arthur C. Clarke had rather extravagantly to say on the subject of Space flight in 1951. He opined that its advent "will produce an expansion of scientific knowledge perhaps unparalleled in history" (p. 153).[8] Yet at the same time, he conceded that "It would be far easier to make the Antarctic bloom like the rose than to establish large self-supporting colonies on such voids as Mars, Ganymede and Titan" (p. 185), the last two bodies here cited being moons of Jupiter and Saturn respectively. Melting much of the south polar ice-cap to make way for a floral landscape would cause the Mean Sea Level of the "indivisible world ocean" to rise scores of metres. In other words, it would be mission impossible.

Reciprocal Dependence

So the conclusion of this analysis is that a reasonably effective screen against natural disasters (and NEO impacts in particular) can gradually be built up: this process being best led, at least for some time, by the permanent members of the UN Security Council, their being backed or cautioned by other states and regional institutions as and where appropriate.

The active operational mission of a Cosmic Command screen should help to give it credibility as a binding force for a new world political order. And such an overarching regime should assist the political geography of our Earth's surface to evolve smoothly. It will be reciprocal dependence. It could be expected to work tolerably well, once all is in place sufficiently. That popular interest in NEO interception in particular may be starting to build up is indicated by the attention being shown in the media in the expectation that asteroid Bennu will make a passage between Earth and Moon in 2035.[9]

Appendix A
The Geostrategic Structure

May one focus briefly on the concept we know as "continental drift". This was enunciated in 1915 by Alfred Wegener (1880–1930), a German physicist valiant in field work and as a theorist, not to mention upfront military experience (1914–15) – this until casualty evacuated.

Wegener was a champion balloonist, outstanding as a polar explorer, and an accomplished meteorologist. Even in English translation, the first major rendering of his "continental drift" hypothesis reads not just definitively but beautifully.[1] Essentially, it says that the Earth's continental crust is always moving, very slowly but persistently, on a viscous, basaltic sea floor. It may be either splitting or coming back together.

The evidence comes from geometry, rock structures, magnetic imprints, climate residuals, palaeontology and so on. A compelling instance of several of these criteria indicating a drifting apart is how well Brazil could once have fitted into the Gulf of Guinea.

Acceptance of the said hypothesis came after a long and dour struggle. But by the sixties of the last century, the drift notion was gaining support from the related field of plate tectonics. Soon the wags would be telling how, before 1965, you could not get a chair in geology unless you denied continental drift; and afterwards, you could not unless you extolled it. Hyperbolic no doubt but expressive of a modicum of truth. Today the notion of continental drift holds sway supremely.

Nodal Western Europe

A 1990 rendering of the main aspects of the sequence, as it has determined the global configuration of today, was as follows. Some 300 million years ago, extremely gradual but sustained movements took place on the part of the two great continents then in being (Laurasia well to the North; and Gondwanaland to the South). These movements caused them so to come together as to create what we know as Pangaea, alias "all lands".[2] What to us is West Eurasia was very much at the centre of a slow grinding collision. This it can still be said to be.

The geographical consequences for the Western Europe we know remain evident and significant. Acute rock folding has left us with very conspicuous mountain chains (most notably the Alps) as well as strongly indented coastlines, proffering valuable ports and fishing grounds. Folding has also preserved coal, metal and salt deposits, quite well dispersed and often accessible for mining. Limestone and sandstone deposits (good for soil development and as building material) are created in shallow seas.

All this has favoured the emergence, through this last millennium, of the European matrix of nation states and statelets, most of them accessing the open ocean. Correspondingly, this ambience has encouraged cultural pluralism along with individualism in enterprise and thought.

The nodality currently conferred on Western Europe is amply confirmed by its presence at the centre of what is known as the "land hemisphere": that half of the Earth's surface which is rather more land than sea, the other hemisphere being very largely sea. What we term the "pole" (i.e. the centre point) of the land hemisphere can be set four square in Brittany. Thus is clinched the truism that Western Europe can access and be accessed very readily for peace or war.

Christendom Central

Christianity was poised at its very outset to take full advantage of this situation. For it had itself arisen just within the eastern bounds of the Roman Empire, well placed to extend early on around and beyond the Mediterranean basin. However, this circumstance told against the "turn the other cheek" passivity of Christ's Sermon on the Mount. It invited rather a spread of more forceful pitches. In due course, the civilizations of the Western Hemisphere were to collapse readily before these, most immediately on account of ignorance of gunpowder. Islam was proving to be a more resilient and reactive adversary. Much the same went for more distant Asian philosophies.

Not that too trite a determinism is warranted. After all, China developed economically during the Middle Ages more strongly than Europe did. However, the Chinese state was then too confined to expand beyond well-defined territorial limits. In rather different ways, its time may now be coming. But the next several decades, the crying need will be for the West and the Islamic world to get on better terms, this based on democratic progress, especially through *dār al-Islām*.

Appendix B
Earthquakes and Volcanoes

There are two terrestrial phenomena which can sometimes have effects akin in scale and character to the impact of a large NEO; and are therefore amenable to similar passive defence. They are earthquakes and volcanoes. The latter mode might suddenly cause destruction sufficiently widespread to put at risk the survival of Humanity or, indeed, of any Earthly life.

The two modes tend to be treated separately in educational textbooks. Yet each is a product of *tectonics* – a process of dynamic deformation taking place inside the Earth's crust. Each category occurs mainly within the mountain-building zone emergent from the Tertiary period (65 to two million years ago). This extends through the Mediterranean and along the Himalayas, then around the Pacific "ring of fire", Sumatra and Honshu being especially susceptible to acute instability.

(1) Earthquakes

Looking at even a small-scale topographical map of the world, there is one submarine feature which stands out singularly. It is the large and deep Nankai trough in the Pacific close under Tokyo Bay, a troughing liable to give rise to major earthquakes all too frequently. The Suruga subduction zone beneath the Nankai trough produced major quakes in 1633, 1703, 1782, 1854, 1923 and 2011 – a fairly regular 74-year average cycle.

The Great Kanto earthquake which struck Tokyo in September 1923 caused 140,000 deaths as well as damage to the tune of 50 billion dollars at 1990 prices. The slowing down of the Japanese economy which resulted was to leave the country and therefore the wider world that much more vulnerable to the dire political consequences of the global economic "blizzard", 1929–32.[1]

In 2005, a report to the Japanese government made what one could say was an obvious recommendation. Move the seat of government from Tokyo. Yet one has heard nothing on that particular score of late. Several reasons might be adduced. With modern electronic means of data dissemination and storage, one could expect another big Tokyo quake to cause less instant chaos, at least in proportion to total activity. Besides which, technical advances have been applied to making buildings able to withstand more shock, while in Tokyo, in particular, there has been a huge investment in a public utility matrix set deep underground. But a big earthquake directly under central Tokyo might still cut Japan's Gross Domestic Product a good while by as much as 40 per cent.[2]

More generally, one has further to bear in mind how vast a literature on earthquake prediction has accumulated the past half century or so. Yet really helpful prognosis, specific and timely, remains "an unrealizable goal for the foreseeable future; and may well be inherently impossible".[3] For one thing, over half a century after its general acceptance in principle, plate tectonics "especially in relation to the deeper Earth, is still poorly understood."[4]

So strategies of quake mitigation, around Tokyo and more generally, may still have to depend too much on probabilistic prediction: an assessment of the mean risk over time apparently inherent in a particular situation for whatever reason.[5] Still, research has lately been indicating that incipient earthquakes may subtly influence local gravitational, magnetic and electrical fields through the ground and the atmosphere.

Comparable work on electrical fields has already been applied to the forecasting of thunderstorms and, indeed, of tropical revolving storms. An indication of the sophistication of current work in this general area is that an ability to measure variations in ambient magnetism to within a billionth of the norm is being sought for the Earth's magnetic field.[6] Meanwhile the mechanics of sea quakes has lately come to be better understood. They sometimes cause oceanic tidal waves, *tsunamis*, across hundreds of miles.

Tsunamis tend to be regarded as the most menacing kind of seismic event. The worst recent instance has been the Sumatran one of 2004. Inexcusable was the lack of an early warning network in states littoral to the Bay of Bengal and other nearby waters. Questions were also asked about the general absence of contingency planning; and the inchoate and overly hyped inflow of international aid.[7] Altogether one has to conclude that tens of thousands of lives were needlessly lost. In Indonesia alone, 170,000 died altogether.

Within several years, however, substantial progress had been made there towards a tsunami alert regime.[8] And sometimes this grim experience had actually done some good for community relations. Witness the peace agreement the government in Djakarta signed with the Free Aceh movement in North Sumatra in May 2005. In Sri Lanka, however, things went the other way.[9]

Even if an earthquake on the 1923 scale, in Tokyo Bay or wherever, proved less traumatic nationally now than was the case then, it could still be a lot more disastrous for the world at large, given the extent of the international economic integration effected in the interim. Indeed, in his capacity as Professor of Geophysical Hazards at University College London, Bill McGwire expressed in 2002 the view that such an event's effects would be "far, far worse" than 1923, nationally as well as globally.[10]

As regards Japan's perceived *immobilisme* over the Tokyo as capital question, there are several points to make in mitigation. One is that very severe earthquakes occurred in the Tokai marine area (near Shizuola, between Tokyo and Nagoya) in 1707 and 1854; and much expert opinion was long persuaded that these events were part of a 120-year average cycle in middle Honshu, making this otherwise nodal area unsuitable for metropolis relocation. Then again, in the twentieth century, Istanbul was hit by numerous quakes. Yet the population of this, Turkey's main city, was left to grow from one to eight million between 1950 and 1990.

In 2016, the authorities in Tokyo promulgated a plan for dealing with a major quake under their city centre. Over 200,000 aid workers can now be deployed within a couple of days or so. But in Tokyo, as in many other places worldwide, the big danger always is that any ill-founded prediction of disaster could lead forthwith to mass panic followed, next time round, by studied indifference.

(2) Volcanoes

Volcanic eruptions can be quite local events, doing not much except exude a modicum of lava. Yet occasionally one will prove truly awesome with extended planetary consequences. This may particularly apply when the force and orientation of a major blast is such as to project debris into the stratosphere where there is no rain to wash the finer fraction downwards. Much may stay aloft seven years or more, encompassing the whole world in a cooling pall throughout.

The oft remarked 1980 eruption of Mount St Helens in Washington

State, USA, is now serving as something of a benchmark for the amount of material said to have been ejected. In the famous 1883 event, Krakatoa (in what we now know as Indonesia) dispatched skywards a good 50 times what Mount St Helens registered in 1980. Similarly, at Tambora (also in Indonesia) an eruption in 1815 expelled an estimated thirty times the Mount St Helens quantity. This led, in Britain and elsewhere, to "winter without a summer", hardly a good augury for post-Napoleonic peace.

The long Little Ice Age, several centuries of climate cooling essentially brought upon the world by the Milankovitch cycles in the Earth's revolution round the Sun and in the former's angle of rotation, reached its nadir between 1693 and 1710.[11] After then, coal-burning supervened. The exact timing of the bottoming out was, however, determined by the extra cooling consequent upon strong volcanic eruptions, e.g. Hekla and Serua in 1693; Amboina in 1694 . . .

The bad harvests duly visited upon *inter alia* northern Europe in the 1690s did have consequences geopolitically. Thus a weakening of the Scottish economy left Edinburgh more prepared to accept, however dourly, the political union with London enacted in 1707. Then led by David Hume (1711–76), the Scottish intelligentsia addressed the attendant identity crisis though less through Gaelic romanticism than by contributing powerfully to the scientific rationalism of the European enlightenment.

A third of the population of Finland had succumbed to epidemics or simply starvation in the "famine year" of 1696–7. The upshot across the Baltic had been the organization against the Swedish empire of a coalition of containment extending from Norway/Denmark round to Russia. Duly, the Great Northern War broke out in 1700. Sweden finally lost the initiative in 1709 after the same bitter winter as had driven an encompassed France to make preliminary peace overtures during the War of the Spanish Succession.

Thira

The cataclysmic eruption of the volcano Thira on the Aegean island of Santorini (now dated at 3643 BP[12]) caused tsunamis across the eastern Mediterranean; and spread a pall of dust around the northern hemisphere. Not long thereafter, the Minoan civilization on Crete went into irregular decline, its absorption into the Mycenean orbit no doubt being part of this recession. In China, the bronze culture Shang dynasty replaced the Neolithic (i.e. New Stone Age), the timing likely being in some measure cause and effect.[13]

What then if one does accept that these particular developments are fairly direct consequences of the Thira vulcanism? Are indirect effects also to be discerned within the comparatively close-knit civilization that by then had arisen across the Eastern Mediterranean and Near East? The Hittites were among several kingdoms which collapsed "without apparent adequate reason". Refugees surged onto the Mediterranean's eastern shores. The palaces of Mycenae and Crete were sacked. Egypt fell "into helpless apathy" for several centuries, hard upon the "nine plagues" and the Hebrew exodus. All in all, around 3200 BP, said Rhys Carpenter (a Classical Archaeologist at Bryn Mawr), "Mediterranean Man has begun to suffer the most severe cultural recession which history records or archaeology can determine".[14] Any connection involves a long reach from Thira in time and place though maybe not an inconceivable one.

Toba

But for Toba in Sumatra *c.* 74,000 BP the eruption ratio as against St Helen's is put at a thousand times. This earlier event has therefore to be seen as exemplifying that most singular of tectonic upheavals, a "supervolcano": a feature in which the volcanic epicentre is underlain by an exceptionally wide and fairly horizontal area of molten rock, *magma*. The *caldera* or depression left by that Sumatran upheaval is today largely filled by a roughly elliptical lake, 85 by 25 kilometres each way. The resultant dust pall will have sustained an acute volcanic winter, more or less globally, for a good ten years.

Before the recent rise to prominence of man-made global warming, one could always say that fluctuations, across intervals of tens of thousands of years, were determined very largely by the Milankovitch cycles. Against this background, a measure of the Toba effect can be gleaned from a seabed core extracted from mid-Atlantic at 53°N. This registers a summer seabed temperature of 14°C in 75,000 BP, falling to 6°C around 72,000 BP.[15]

The Toba pall was apparently especially pronounced over India, causing high mortality within our species as well as many others. The subcontinent was to be peopled anew from both West and East. So far as the Eastern Hemisphere as a whole was concerned, our population may have troughed out at something like 50 to 100 thousand by then, having entered Eurasia and become well spread in general. We had yet to make it into the Western Hemisphere.

Many particular human communities will have died out. In the Kruger National Park in South Africa in 1986, one was told that 400 was seen as a representative minimum for the healthy survival of a detached community of largish mammals. A similar figure might apply to an inbred English village, pre-1939.

Most remarked as a potential site for a supervolcanic eruption is the Yellowstone National Park in Wyoming. Its hot springs and geysers are set within a large *caldera* formed by three huge eruptions between 500,000 and 2,500,000 years ago. Similarly large structures are found in California and New Zealand as well as Indonesia. Lake Taipo in New Zealand produced the most recent super-eruption, this in 26,000 BP. Supervolcanoes may otherwise erupt in locales not currently suspect in this regard.

Notes

Prefatory Statement

1 *BL*, pp. 249–52.
2 Samuel Glasstone and Philip J. Dolan, *The Effects of Nuclear Weapons* (Washington DC: US Government Printing Office, 1977.
3 A meteorite is a meteor that has already struck the Earth.
4 *The Guardian*, 13 December 2016.
5 Theodore P. Snow, *The Dynamic Universe* (St. Paul: West Publishing, 1991), pp. 282–6.
6 Margaret Thatcher, *The Downing Street Years* (London: HarperCollins, 1993), p. 467.

The Author

1 Neville Brown (ed.), *American Missile Defence. Views from China and Europe*, Current Decisions Report 25 (Oxford: Oxford Research Group, May 2000).

1 Legends and Realities

1 H. G. Wells, *The Shape of Things to Come* (London: Hutchinson, 1933), pp. 202–19.
2 See the author's *The Future Global Challenge* (New York: Crane Russak, 1977), pp. 49–50.
3 Mike Ashley, *Out of This World* (London: The British Library, 2011), pp. 116–17.
4 William E. Burrows, *The Asteroid Threat* (New York: Prometheus Books, 2014), Chapter 4.
5 Neville Brown, *New Strategy Through Space* (Leicester: Leicester University Press, 1990), Chapter 4.
6 Immanuel Velikovsky, *Worlds in Collision* (London: Macmillan, 1950), p. 99.
7 E. C. Krupp (ed.), *In Search of Ancient Astronomers* (London: Doubleday & Co., 1979), p. 256.
8 John Gribbin and Stephen Plagemann, *The Jupiter Effect* (London: MacDonald, 1974), p. 116.
9 Neville Brown, *The Fundamental Issues Study. Within the British BMD Review* (Oxford: Mansfield College, February 1998), p. 137.

10 "China seeks atom option to fend off asteroids", *International Herald Tribune*, 21 April 1996.
11 William James, *The Varieties of Religious Experience* (New York: Longman Greens, 1925), p. 367.
12 Stuart Greenstreet, "Wittgenstein, Tolstoy and the Folly of Logical Positivism", *Philosophy Now* 103 (July/August 2014): 6–11.
13 Zbigniew Brzezinski, "The International and the Planetary", *Encounter* XXXIX, 2 (August 1972): 49–55.
14 Patricia Mishe, *Star Wars and the State of Our Souls* (Minneapolis: Winston Press, 1985), p. 52.

2 Galactic Reconnaissance

1 Gerard K. O'Neill, *The High Frontier. Human Colonies in Space* (New York: William Morrow, 1977), Chapter XII.
2 Freeman Dyson, *Disturbing the Universe* (NewYork: Harper and Row, 1979), Chapter 21.
3 *Ibid.*, p. 124.
4 Martin Rees, "Special Report: the Science of Eternity", *Prospect* 70 (January 2002): 50–4.
5 See Kevin Heng and Joshua Winn, "The next great exploration hunt", *American Scientist* 103, 3 (May–June 2015): 196–207.
6 Keith Cooper, "The largest telescopes in the world", *Astronomy Now* (April 2010): 56–67.
7 Nicholas Wright, "Star Formation in the Gaia Era", *Astronomy and Geophysics* 57, (February 2016): 1.37–1.43.
8 Kate Dutton, "Opening a Window on the Universe", *Astronomy and Geophysics* 56 (April 2015):2.22–2.26.
9 George Dyson, *Project Orion* (London: Allen Lane, 2002), Chapter 18.
10 Peter E. Glazer, "Evolution of the Solar Power Satellite concept: the utilisation of energy from Space", *Space Solar Power Review* 4 (1983): 11–21.
11 Jeffrey Mazo, "Geoengineering: Don't Just Do Something, Stand There", *Survival* 57, 3 (June/July 2015): 185–92.
12 Donald Y. Yeomans, *Near Earth Objects* (Princeton: Princeton University Press, 2013), Chapter 9.
13 Peter Marsh, *The Space Business* (Harmondsworth: Penguin, 1985), Chapter 7.
14 V. C. Avduyevsky (ed.), *Manufacturing in Space: Problems and Advances* (Moscow, 1985), pp. 155–6.
15 *The Economist*, 9 April 2016.
16 Richard Branson, "From the Mojave to the Moon" in Daniel Franklin (ed.), *The World in 2014* (London: *The Economist*, 2013), p. 62.
17 B. H. Liddell-Hart (ed.), *The Soviet Army* (London: Weidenfeld and Nicolson, 1956), Chapter 21.
18 *GHC*, p. 322.
19 Sam Jones, "Satellite Wars", *Financial Times Commercial Magazine*, November 21/22, 2015.

20 John H. Barton and Lawrence B. Weiler, *International Arms Control* (Stanford: Stanford University Press,1976), pp. 76–8.
21 For the full text, see J. E. S. Fawcett, *Outer Space: New Challenges to Law and Policy* (Oxford: Clarendon Press, 1984), Appendix B.
22 David Whitehouse, "Law in Orbit", *The New Scientist* 119, 1620 (7 July 1988): 41–43.

3 Tough Centuries Ahead

1 Michael Potter in Joseph N. Pelton and Firooz Allahdadi (eds.), *Handbook of Cosmic Hazards and Planetary Defense* (Heidelberg, Springer Reference 2015 (2 Vols.), Vol. 2, p. 1049.
2 Bill McGuire, *Global Catastrophes* (Oxford: Oxford University Press, 2014), p. 101.
3 *BL*, p. 201.
4 Eli Pariser, *The Filter Bubble* (London: Viking, 2011), p. 243.
5 Susan Greenfield, *The Quest for Identity in the 21st Century* (London: Sceptre, 2008).
6 Miguel Nicholeus, "Actions from Thoughts", *Nature* 409 (18 January 2001): 403–7.
7 W. E. Woodson and D. W. Conover, *Human Engineering Guide* (Berkeley: University of California Press, 1964), pp. 6–20.
8 Duncan Graham-Rowe, "Mind Readers", *New Scientist* (28 May 2011): 40–3.
9 "Print me a Stradivarius", *Economist*, 12 February 2012.
10 Edward Luttwak, "Why Fascism is the Wave of the Future", *London Review of Books* 16, 7 (7 April 1994): 3–6.
11 "Meet your Unborn Child before it's Conceived", *New Scientist* (12 April 2014): 15–19.
12 Harouf A. Hasian, *The Rhetoric of Eugenics in Anglo-American Thought* (Athens: University of Georgia Press, 1996), Chapter 5.
13 Dr Nadia Sadak, *The State of World Population, 1988* (New York: UNPF, 1988), Table 1.
14 *Bulletin of the Atomic Scientists* 62, 2 (March–April 2006): 34.
15 "Building the Dream", *Special Report on China, Economist* (19 April 2014), p. 4.
16 *Ibid.*, p. 7.
17 *Special Report*, "The World Goes to Town", *Economist* (5 May 2007): 13.
18 Ross Adams, "Looking for a greener present. Neo-liberalism and the Ecocity", *Radical Philosophy* 163 (September–October 2010): 2–7.
19 Edward Gleaser, *The Triumph of the City* (London: Pan Macmillan, 2011), pp. 268–9.
20 "Building the Dream", *Economist*, (19 April 2014), p. 11.
21 H. H. Lamb, *Climate, History and the Modern World* (London: Routledge, 1995), p. 179.

22 Neville Brown, *History and Climate Change, a Eurocentric Perspective* (London: Routledge, 2001), p. 68.
23 *Ibid.*, pp. 61–3.
24 *Ibid.*, p. 296.
25 *Ibid.*, p. 139.
26 K. Pribyl, "The study of the climate of medieval England: a review of historical climatology's past achievements and future potential", *Weather* 69, 5 (May 2014): 116–20.
27 For a discussion of this contention see "the Dantean anomaly" in Neville Brown, *History and Climate Change, a Eurocentric Perspective* (London: Routledge, 2001), pp. 251–4.
28 Brown, *History and Climate Change*, pp. 214–17.
29 "Impacts of Climate Change", *Weather* 69, 5 (May 2014): 114.
30 *The Kodanska Encyclopedia of Japan* (9 Vols.), (Tokyo: Kodansha, 1983), Vol.8, p. 122.
31 J. Neumann, "Great Historical Events that were Significantly Affected by the Weather", 1. The Mongol Invasions of Japan", *Bulletin of the American Meteorological Society* 56, 11 (November 1975): 1167–71.
32 *Weather* 17, 12 (December 2015): 334.
33 *BL*, Chapter 1(2).
34 *GHC*, pp. 135–7.
35 Mark Bassin, "Inventing Siberia: visions of the Russian East in the early 19th century", *American Historical Review* 96, 3 (June 1991): 763–94.
36 Reprinted in *Wildlife*, March 1976.
37 "Pumping Heat", *Economist Technology Quarterly*, 7 June 2014, p. 5.
38 Mancur Olson, "Rapid Growth as a Destabilising Force", *Journal of Economic History* XXIII, 4 (December 1963): 529–32.
39 *BL*, pp. 110–11.
40 Daniel Freeman and John Freeman, *Paranoia, The Twenty-First Century Feat* (Oxford: Oxford University Press, 2008), Chapter 3.

4 Planetary Confinement

1 Samuel Charap, "The Ghost of Hybrid War", *Survival* 57, 6 (December 2015): 51–5.
2 See the author's "The Jordanian Civil War", *The Military Review* 61, 9 (September 1971): 38–48.
3 René Albrecht-Carrie, *The Meaning of the First World War* (Prentice Hall: Englewood Cliffs, 1965), pp. 42–3.
4 Peary's contested claim was to be brilliantly corroborated by the 2005 Anglo-Canadian expedition. See Tom Avery, *To the End of the Earth* (New York: St. Martin's Press, 2009).
5 J. E. Black, "The Canadian Pacific Railway, as a Model for the Trans-Siberian Railway", *Sibirica*, 4, 2 (October 2004): 196–200.
6 *BL*, pp. 26–9.

7 Bertrand Russell, *A History of Western Philosophy* (London: Unwin University Press, Undated *c.* 1942), p. 581.
8 *EC*, p. 57.
9 David G. Chandler, *The Campaigns of Napoleon* (New York, Macmillan, 1966).
10 Andrew Hurrell, "Hegemony, liberalism and global order: what space for would-be powers?", *International Affairs* 82, 1 (January 2006): 1–19.
11 *EC*, pp. 222–26.
12 Albertine Gaur, *A History of Writing* (London: The British Library, 1992), pp. 106–17.
13 "An uncertain community", *The Economist*, 29 October 2016.

5 The Abrahamic Schism

1 E.g. A. C. Bouquet, *Comparative Religion* (Harmondsworth: Penguin, 1971), p. 277.
2 David Keys, *Catastrophe* (London: Century Books, 1999), pp. 61–4.
3 Abdullah Yusuf Ali (ed.), *The Holy Qur'an* (Elmhurst Tahrike: Tarsile Qur'an, 1987), *sūra* XXXIV, v. 16. The sacred text is divided into *sūra* or thematic chapters.
4 Neville Brown, *History and Climate Change* (London: Taylor and Francis, 2001), p. 111.
5 *EC*, pp. 198–9.
6 Fred McGraw Donner, "Mecca's Food Supplies and Muhammad's Boycott", *Journal of the Economic and Social History of the Orient* XX, III (1997): 249–66.
7 The Department of Geography, King Saud University, *Weather*, 71,10 (October 2016): 266.
8 Xavier de Planhal, *The World of Islam* (Ithaca: Cornell University Press, 1959), pp. 2–8.
9 R. M. Holt *et al.* (eds.), *The Cambridge History of Islam*, 2 Vols (Cambridge: Cambridge University Press, 1970), Vol. 2, p. 445.
10 Ninian Smart, *The World's Religions* (Cambridge: Cambridge University Press, 1998, p. 289.
11 Ali (ed.), *The Holy Qur'an*, note 204.
12 Ronald L. Nettler, *Sūfi Metaphysics and Qur'anic Prophets* (Cambridge: Islamic Texts, Society, 2003), p. 17.
13 Seyyed Hossein Nasr, *Ideals and Realities of Islam* (London: George Allen and Unwin, 1966), p. 16.
14 John Glubb, *A Short History of the Arab Peoples* (London: Hodder and Stoughton, 1969), pp. 104–5.
15 Brown, *History of Climate Change*, pp. 113–18.
16 Nasr, *Ideals and Realities of Islam*, pp. 6–7.
17 Seyyed Hossein Nasr, *Islamic Science* (Westerham: World of Islam Festival, 1976), pp. 17–20.
18 Fazlur Rahman, *Islam* (London: Weidenfeld and Nicolson, 1966), p. 186.

19 AH means After Haj, the initiation by the Prophet himself of the annual pilgrimage to Mecca. Every Moslem who can afford it is expected to participate once in a lifetime in what is now an annual mass event.
20 S. Pines, "What was original in Arabic Science" in A. C. Crombie (ed.), *Scientific Change* (New York: Basic Books, 1963), Chapter 6.
21 Seyyed Hossein Nasr, *An Introduction to Islamic Cosmological Doctrines* (London: Thames and Hudson, 1978), p. 165.
22 Bertrand Russell, *History of Western Philosophy* (London: Unwin, undated), p. 417.
23 John North, *Astronomy and Cosmology* (London: Fontana History of Science, 1994), pp. 261 and 273.
24 Akbar S. Ahmed, *Living Islam* (London: BBC Books, 1993), p. 77.
25 *BL, p.* 228.
26 Carrie Rosefsky Wickham, *The Moslem Brotherhood* (Princeton: Princeton University Press, 2013), pp. 27–29.
27 Zakir Naik. *The Qur'an and Modern Science* (Riyadh, 2007).
28 David A. King, "Mathematical Astronomy in Islamic Civilisation", in Helaine Selin (ed.), *Astronomy Across Cultures* (Dordrecht: Kluwer, 2000), pp. 585–613.
29 S. Pines in A. C. Crombie (ed.), *Scientific Change* (NewYork: Basic Books, 1963), Chapter 6, "What was Original in Arabic Science".
30 Clinton Bailey, "Bedouin Star-Lore in Sinai and the Negev", *Bulletin of the School of Oriental and African Studies* Vol. XXXVII, Part 3, 1974: 580–596.
31 Oliver Moody, Science Correspondent, *The Times*, 18 October 2016.
32 Roy Jenkins, *Churchill* (London: Pan, 2001), p. 350.
33 Martin Gilbert, *Winston Churchill* (London: Heinemann, 1975), p. 563.
34 *The Guardian*, 4 April 1986.
35 Hannah Arendt, *The Origins of Totalitarian Democracy* (New York: Harcourt Brace, 1958), three parts, Part 1, p. 6.
36 Albert Einstein (Alan Harris, trans.), *The World As I See It* (London: Watts, 1940), pp. 96 and 105.
37 Walter Clay Lowdermilk, *Palestine, Land of Promise* (London: Victor Gollancz, 1944), pp. 55–6.
38 Rhys Carpenter, *Discontinuity in Greek Civilisation* (Cambridge: Cambridge University Press, 1966), pp. 3 and 15–18.
39 Arie S. Issar in Arie S. Issar and Neville Brown (eds.), *Water, Environment and Society in Times of Climate Change* (Dordrecht, Kluwer, 1998), Chapter 6.
40 Amos Frumkin *et al.*, in *Ibid*, Chapter 5.
41 David Watkins, *The Exceptional Conflict* (London: Council for the Advancement of Arab–British Understanding, 1984).
42 e.g. David Shulman, "Gaza: the Murderous Melodrama", *New York Review of Books* LXI, 18 (20 November to 3 December 2014): 32–34.
43 Hannah Arendt in Michael Selzer (ed.), *Zionism Reconsidered* (London: Macmillan, 1979), pp. 213–49.

44 *GHC*, pp. 184–88.
45 Edgar O'Ballance, *The Sinai Campaign* (London: Faber and Faber, 1959), Chapter 2.
46 Avi Shlaim, "Talking to the Enemy", *New Statesman*, 19 May 2008.
47 See the author's "The Jordanian Civil War", *The Military Review* L I, 9 (September 1971): 38–48.
48 Elisha Efrat, *Geography and Politics in Israel since 1967* (London: Frank Cass, 1988), Chapter 3.
49 Interviewed in *Newsweek*, 7 November 2011.

6 The Thermonuclear Five

1 *BL*, pp. 271–2.
2 Arnold Toynbee, *Surviving the Future* (London: Oxford University Press, 1971).
3 Bertrand Russell, *Unpopular Essays* (London: George Allen and Unwin, 1950), Chapter X.
4 "Follow the Trail", *The Economist*, 12 November 2016.
5 *BL*, Appendix A.
6 David Deudney and Hans W. Maull, "How Britain and France could reform the UN Security Council", *Survival* 53, 5 (October–November 2015): 107–8.
7 William Walker, "Trident's replacement and the survival of the United Kingdom", *Survival* 67, 5 (October–November 2015): 7–26.
8 Evgenny Lebedev, "The West would be foolish not to accept Putin's grand design", *Evening Standard*, 29 September 2015.
9 *China Daily*, 2–8 October 2015.
10 *GHC*, Appendix B, "Explosive Nuclear Release".

7 Skyward Precursors

1 David G. Rowley, *Millennarian Bolshevism, 1900–1920* (New York: Garland Publishing, 1987), p. 345.
2 Leon Trotsky, *Literature and Revolution* (Ann Arbor: University of Michigan, 1960, pp. 210–16.
3 A. T. Grigorian in Charles Coulson Gillespie (editor-in-chief) in *Dictionary of Scientific Biography* (New York: Charles Scribner, 1976) (16 volumes), vol. XIII, pp. 483–4.
4 Vladimir Lytkin, Ben Finney and Lindmila Alepka, "Tsiolkovsky, Russian Cosmism and Extra-Terrestrial Intelligence", *Quarterly Journal of the Royal Astronomical Society* 36, 4 (December 1995): 369–74.
5 Terence Armstrong, *The Northern Sea Route* (Cambridge: Cambridge University Press, 1957), pp. 39–40 and 49.
6 Milan L. Hauner, "Stalin's Big Fleet Program", *Naval War College Review* LVII, 2, (Spring 2004): 87–120.
7 In the West, a Medium Range Ballistic Missile (MRBM) was defined early

on as having a range between 600 and 1500 nautical miles. An Intermediate Range Ballistic Missile (IRBM) covered 1500 to 3000 nm. An Intercontinental Ballistic Missile (ICBM) could travel more than 3000 nm, quite possibly several times more.'Intercontinental' really meant 'inter-hemispherical'.

8 Neville Brown, "Towards the Superpower Deadlock", *World Today* XXII, 9 (September 1966): 366–74.
9 Mark Bassin, "Inventing Siberia: Visions of the Russian East in the early 19th century", *American Historical Review* 90, 3 (June 1991): 764–94.
10 Neville Brown, *History and Climate Change* (London: Routledge, 2001), p. 139.
11 E. Royston Pike, *Ethics of the Great Religions* (London: Watts, 1948), p. 215.
12 *EC*, pp. 227–8.
13 Joseph Needham, *Science and Civilization in China*, 23 vols. to date (Cambridge: Cambridge University Press, 1959–2004), Vol. 3, pp. 459–61.
14 Robert K. G. Temple, *China: Land of Discovery* (Wellingborough: Stephens, 1986), pp. 33–4.
15 Richard Stephenson and Kevin Yau, "Oriental tales of Halley's Comet", *The New Scientist* 103, 1423 (24 September 1984): 30–32.
16 Xu Zhentao, "Solar observations in ancient China and solar variability", *Philosophical Transactions of the Royal Society of London*, A 330, 1990, pp. 513–15.
17 C-Ching Chu, "Climate Pulsations during Historic Time in China", *The Geographical Review* XV1, 2 (April 1926): 274–9.
18 Needham, Table 39.
19 Brown, *History and Climate Change*, p. 69.
20 Tu-cheng Yeh *et al.*, "An international symposium on climate in Beijing . . .", *Bulletin of the American Meteorological Society* 66, 9 (September 1985): 1147–52.
21 William H. McNeill, *The Pursuit of Power* (Oxford: Basil Blackwell, 1983), p. 50.
22 Sally and David Dugan, *The Day the World Took Off* (Basingstoke: Macmillan, 2000), pp. 128–33.
23 Brown, *History and Climate Change*, p. 212.
24 Gavin Menzies, *1421* (London: Bantam Press, 2002), pp. 323–30.
25 Nayan Chandra, "Sailing into Oblivion", *Far Eastern Economic Review* 162, 36 (9 September 1999): 44–6.
26 H. H. Gerth and C. Wright Mills (ed.), *From Max Weber* (London: Routledge and Kegan Paul, 1970), Chapter XVII, Sections 5 to 7.
27 Wang Juntao in Daniel A. Bell and Hahm Chaibong (eds.), *Confucianism for the Modern World* (Cambridge: Cambridge University Press, 2003), Chapter 3.
28 Ian Barbour, *Religion in an Age of Science,* 2 vols. (New York: Harper San Francisco, 1991), Vol. 1, p. 119.

29 Fritjof Capra, *The Tao of Physics* (New York: HarperCollins, 1991), p. 30.
30 *Ibid.,* p. 93.
31 *EC*, pp. 256–7.
32 Bertrand Russell, *Mysticism and Logic* (London: Longmans and Green, 1918), p. 8.
33 *Ibid.,* p. 12.
34 Neville Brown (ed.), *American Missile Defence: Views from China and Europe* (Oxford: Oxford Research Group, 2000), 77.
35 E.g. Edward Thompson, "Notes on Exterminism: the Last Stage of Civilisation, *New Left Review* 121 (May–June 1980): 3–31.
36 John North, *Astronomy and Cosmology* (Hammersmith, Fontana, 1994), p. 133.
37 Helen Fitzwilliam, "Chinese foreign legion", *The World Today* 70, 3 (June and July 2015: 32–6.
38 See Tariq Ali as cited in Chapter 6(3).
39 Neville Brown, "The myth of an Asian diversion", *Journal of the Royal United Services Institute* 118, 3 (September 1973): 48–51.
40 "Xi's New Model Army", *The Economist*, 16 January 2016.
41 Joan Johnson-Freeze, *Space as a Strategic Asset* (New York: Columbia University Press, 2000), p. 197.
42 *The Military Balance 2016* (London: International Institute for Strategic Studies, 2016), pp. 241, 189 and 88.
43 Jeffrey Lewis, *Paper Tigers. China's Nuclear Posture* (Abingdon: Routledge for International Institute for Strategic Studies, 5 December 2014).
44 Schrödinger's panda", *The Economist*, 4 June 2016.
45 Nigel Inkster, *China's Cyber Power* (London: International Institute for Strategic Studies, 2016), p. 19.

8 Terrestrial Belligerence

1 Neville Brown, *The Future Global Challenge* (New York: Crane Russak for RUSI), 1977, p. 297.
2 Quoted in Sam Jones, "Satellite Wars", *Financial Times.com Magazine*, November 21/22, 2015.
3 Anne Savage (trans.), *The Anglo-Saxon Chronicle* (London: Phoebe Phillips, 1982), p. 73.
4 Dr Norman Friedman, "UCAVs: A New Kind of Air Power", *Royal United Services Institute, Defence Systems* (June 2010): 62–4.
5 "The see-through sea", *The Economist*, 16 July 2016.
6 "Ghost Ships", *Economic Technology Quarterly*, 8 March 2014, p. 3.
7 *BL*, Appendix C.
8 Alessio Patalona, "Beyond the Gunboats, Rethinking Naval Diplomacy and Humanitarian Assistance. Disaster Relief in East Asia", *RUSI Journal* 160, 2 (April/May 2015): 32–9.
9 Neville Brown, *History and Climate Change. A Eurocentric Perspective* (London: Routledge, 2001), pp. 268–9.

10 Quoted in George Dyson, *Project Orion* (London: Allen Lane, 2002), pp. 276–7.
11 Jean Guillemin, "Seduced by the State", *Bulletin of the Atomic Scientists* 63, 5 (September/October 2007): 14–16.
12 See the author's *Fundamental Issues Study. Within the British Ballistic Missile Defence Review* (Oxford: Mansfield College, February 1998), p. 66.
13 E.g. Robert L. Dorit, "Breached Ecological Barriers and the Ebola Outbreak", *American Scientist* 103, 4 (July–August 2015): 256–59.
14 Peng Wang and Stephen Blancke, "Mafia State. The Evolving Threat of North Korean Narcotics Trafficking", *Royal United Services Institute Journal* 159, 5 (October/November 2014): 57–59.
15 Malcolm Dando, *Biological Warfare in the 21st Century* (London: Brassey's, 1994), p. 98.
16 John Leslie, "The risk that human beings will soon be extinct", *Philosophy* 85, 335 (October 2010): 447–63.
17 Wendy Barnaby, *The Plague Makers* (New York: Continuum, 1999), p. 334.
18 Jungmin Kang *et al.*, "South Korea's Nuclear Surprise", *Bulletin of the Atomic Scientists* 61, 1 (January/February 2005): 46–49.
19 International Thermonuclear Experimental Reactor based in France.

9 Near Earth Objects

1 Donald K. Yeomans, *Near Earth Objects* (Princeton: Princeton University Press, 2013), Table 1.1.
2 Carl Sagan, "Nuclear War and Climate Catastrophe: Some Policy Implications", *Foreign Affairs* 62, 1 (Winter 1983–4): 257–94.
3 Sergei Kapitza, "A Soviet View of Nuclear Winter", *Bulletin of the Atomic Scientists* 41, 9 (October 1985): 37–9.
4 *Report of the Task Force on Potentially Hazardous Near Earth Objects* (London: British National Space Centre, September 2000), Annex B, Table B1.
5 Victor Clube and Bill Napier, *The Cosmic Winter* (Oxford: Basil Blackwell, 1990), p. 158.
6 Mark Boslough and colleagues noted in Yeomans, *op. cit.,* p. 119.
7 *Report of the Task Force,* p. 16.
8 William E. Burrows, *The Asteroid Threat* (Amherst, Prometheus Books, 2014), p. 38.
9 Karel A. van de Johannes Anderson in Joseph N. Pelton and Firooz Allahdadi (ed.), *Handbook of Cosmic Hazards and Planetary Defense* (Heidelberg: Springer Reference, 2015), Vol. 2, p. 757.
10 Theodore P. Snow, *The Dynamic Universe* (St. Paul: West Publishing Company, 1991), p. 336.
11 A. Lewis Licht, "The Rate of Naked Eye Comets from 100 BC to 1970 AD", *Icarus* 137, 2 (February 1999): 355–6.

12 L. E. Alvarez *et al.*, "Extra-terrestrial cause for the Cretaceous-Tertiary extinction", *Science* 208 (1980): 1095–1108 (?).
13 Clube and Napier, p. 141.
14 Catherine Hills, "Roman Britain to Anglo-Saxon England", *History Today* 40, 10 (October 1990): 46–52.
15 Michael Winterbottom (ed.), *Gildas, the Ruin of Britain and Other Works* (London: Phillimore, 1978), pp. 16–17.
16 *Ibid.,* p. 13.
17 *Ibid.,* p. 27.
18 Clube and Napier, pp. 107–10.
19 K. Rotherford Davies, *Britons and Saxons, the Chiltern region, 400–700* (Chichester: Phillimore, 1982), p. 9.
20 E.g. Hector Munro Chadwick in H. M. Chadwick *et al., Studies in Early Church History* (Cambridge: Cambridge University Press, 1939), p. 21.
21 Article reprinted in English in *Chinese Astronomy,* No. 1, 1977, pp. 177–200.
22 Clube and Napier, pp. 15–16.
23 Theodore Snow, pp. 538–9.

10 Celestial Engagement

1 Joan Johnson-Freeze, *Space as a Strategic Asset* (New York: Columbia University Press, 2007), p. 64.
2 His speech to the New York Economics Club, 18 November 1963.
3 Donald K. Yeomans, *Near-Earth Objects* (Princeton: Princeton University Press, 2013), p. 147.
4 Edward T. Lu and Stanley G. Love, "Gravitational Tractor for Towing Asteroids", *Nature* 435, 7065 (10 November 2005): 177–8.
5 Yeomans, pp. 144–5.
6 L. Drube *et al.*, "Neoshields" in J. N. Felton and F. Allahdadi (eds.), *Handbook of Cosmic Hazards and Planetary Defense (*Heidelberg: Springer International, 2015), 2 Vols., Vol. 2, pp. 763–90.
7 William E. Burrows, *The Asteroid Threat* (Amherst: Prometheus Books, 2010), p. 67.
8 David R. S. Dearborn and Paul L. Miller, "Defending against Asteroids and Comets" in Felton and Allahdadi, *Handbook of Cosmic Hazards and Planetary Defense,* Vol. 2, pp. 733–53.
9 Yeomans, p. 146.
10 *Ibid.*, p. 146.

11 Emergent Priorities

1 Vincent Brome, *H. G. Wells* (London: Longmans Green, 1951), Chapter 14.
2 "Starship Enterprise", *The Economist,* 16 April 2016.

3 Stamatios M. Krimigis and Robert B. Becker, "The Voyager's Odyssey", *American Scientist* 103, 4 (July–August 2015): 284–91.
4 Andra J. Wolfe, "Germs in Space", *Isis* 93, 2 (June 2002): 183–205.
5 *BL*, p. 198.
6 *Ibid.*, pp. 193–5.
7 Richard Cronin, "Mekong Dams and the Perils of Peace", *Survival* 51, 6 (December 2009–January 2010): 147–60.
8 "Essay on the Mekong", *The Economist*, 21 March 2015.
9 Tao Hanshang, *Sun-Tsu, the Art of War* (Ware: Wordsworth Editions, 1993), p. 105.
10 *EC*, pp. 5–8.
11 Stephen Baxter, "Our Future in Space", *Focus* Issue 211 (April 2010): 27–34.
12 Ayers rock is or was an aboriginal shrine in central Australia.
13 Derived from Table 2.12 in Samuel Glasstone and Philip J. Dolan, *The Effects of Nuclear Weapons* (Washington D.C.: US Department of Defense, 1977.
14 "More from Moore", *Economist Technology Quarterly*, 5 September 2015.

12 The Cosmic Setting

1 Roger Saltair, *Pascal: The Man and the Message* (London: Blackie, 1927), p. 197.
2 Arthur B. Evans and Ron Miller, "Jules Verne: Misunderstood Visionary", *Scientific American* 4, 276 (April 1967): 73–81.
3 John North, *Astronomy and Cosmology* (London: Fontana, 1994), pp. 440–1.
4 Tenzin Gyatso, *Ancient Wisdom, Modern World* (London: Little Brown, 1999), Chapter 15.
5 In making this judgement, no aspersions whatsoever are intended against his successor, still very new in post.
6 Jonathan Sacks, *The Dignity of Difference* (London: Continuum, 2002), pp. 274–5.
7 *EC*, p. 275.
8 Arthur C. Clarke, *The Exploration of Space* (London: Temple press, 1951).
9 Sarah Knapton, Science Editor, "Lift-off for NASA mission to save Earth from killer asteroid", *Daily Telegraph*, 10 September 2016.

Appendix A – The Geostrategic Structure

1 Alfred Wegener, John Biram (trans.), *The Origin of the Continents and Oceans* (New York: Dover Publications, 1966). This modern classic first appeared in 1915.
2 Cheryl Sliver and Ruth Defries, *One Earth, One Future* (Washington DC: National Academy Press, 1990), Chapter 3.

Appendix B – Earthquakes and Volcanoes

1 Peter Hadfield, *Sixty Seconds that will change the World* (London: Pan, 1992), Chapter 1.
2 Robin Harding and Steve Barnard, "Is Japan ready for the 'Big One'?", *Financial Times*, 18 May 2016.
3 Francesco Malargia and Robert J. Geller (ed.), *Earthquake Science and Science Risk Prediction* (Dordrecht: Kluwer for NATO Science Series, 50, 2003), pp. 34–7.
4 James Wookey, "New Views of the Earth's Interior", *Astronomy and Geophysics* 50, 3 (June 2009).
5 Seth Stein and Anke Friedrich, "How much can we clear the crystal ball?", *Astronomy and Geophysics* 55 (April 2016): 2.11–2.21.
6 Julian Hunt *et al.* "Tsunamis and geophysical warnings", *Astronomy and Geophysics* 51, 5 (October 2010): 5.37 to 5.38.
7 Benedict Korf, "Disaster, Generosity and the Other", *Geographical Journal* 172, 3 (September 2006): 245–7.
8 Amitav Acharya, *Constructing a Security Community in South-East Asia* (London: Routledge, 2009), p. 252.
9 Jennifer Hyndman, "The Securitization of Fear", *Annals of the Association of American Geographers* 97, 2 (June 2007): 338–49.
10 Bill McGwire, *Global Catastrophes* (Oxford: Oxford University Press, 2002), pp. 93–5.
11 Neville Brown, *History and Climate Change* (London: Routledge, 2001), Chapter 2.
12 BP is Before Present.
13 Mike Baillie, "Do Irish bogs date the Shang dynasty?", *Current Archaeology* X, 16 (November 1980): 310–13.
14 Rhys Carpenter, *Discontinuity in Greek Civilisation* (Cambridge, Cambridge University Press, 1966, pp. 8 and 16–18.
15 Alastair Dawson, *Ice Age Earth* (London: Routledge, 1992), Fig. 10.7.

Index

www.ingramcontent.com/pod-product-compliance
Lightning Source LLC
Chambersburg PA
CBHW071248180926
51613CB00002B/162
* 9 7 8 1 8 4 5 1 9 7 7 0 4 *